Universitext

Komaravolu Chandrasekharan

Classical Fourier Transforms

Springer-Verlag
Berlin Heidelberg New York
London Paris Tokyo

Komaravolu Chandrasekharan
Professor of Mathematics
Eidgenössische Technische Hochschule Zürich
CH-8092 Zürich

Mathematics Subject Classification (1980): 42-XX, 10-XX, 60-XX

ISBN-13: 978-3-540-50248-7 e-ISBN-13: 978-3-642-74029-9
DOI: 10.1007/978-3-642-74029-9

Library of Congress Cataloging-in-Publication Data
Chandrasekharan, K. (Komaravolu), 1920-
Classical Fourier transforms / Komaravolu Chandrasekharan.
p. cm.–(Universitext). Bibliography: p.

1. Fourier transformations. I. Title.
QA403.5.C48 1989 88-38192 515.7'23–dc 19 CIP

2141/3140-543210 – Printed on acid-free paper

Preface

In grateful remembrance of
Marston Morse and John von Neumann

This text formed the basis of an optional course of lectures I gave in German at the Swiss Federal Institute of Technology (ETH), Zürich, during the Wintersemester of 1986-87, to undergraduates whose interests were rather mixed, and who were supposed, in general, to be acquainted with only the rudiments of real and complex analysis. The choice of material and the treatment were linked to that supposition. The idea of publishing this originated with Dr. Joachim Heinze of Springer-Verlag. I have, in response, checked the text once more, and added some notes and references. My warm thanks go to Professor Raghavan Narasimhan and to Dr. Albert Stadler, for their helpful and careful scrutiny of the manuscript, which resulted in the removal of some obscurities, and to Springer-Verlag for their courtesy and cooperation. I have to thank Dr. Stadler also for his assistance with the diagrams and with the proof-reading.

Zürich,
September, 1987

K.C.

Contents

Chapter I. Fourier transforms on $L_1(-\infty, \infty)$

§1. Basic properties and examples

We assume as known Lebesgue's theory of integration.

If p is any real number, with $p \geq 1$, we denote by $L_p(-\infty,\infty)$ the vector space of all complex-valued functions f(x) of the real variable x, $-\infty < x < \infty$, such that f is Lebesgue measurable, and

$$\| f \|_p = \left(\int_{-\infty}^{\infty} |f(x)|^p dx \right)^{1/p} < \infty.$$

We call the number $\| f \|_p$ the L_p-norm of f.

If $f,g \in L_p(-\infty,\infty)$, we say that f is *equivalent* to g, and write $f \equiv g$, if and only if $f = g$ except for a set of Lebesgue measure zero. The relation '$\equiv$' is reflexive, symmetric, and transitive, and partitions $L_p(-\infty,\infty)$ into *equivalence classes*, and $L_p(-\infty,\infty)$ is a *Banach space* if it is looked upon as the set of all such equivalence classes, the *norm* of an equivalence class being defined as the L_p-norm of any of its members.

We shall use the same symbol $L_p(-\infty,\infty)$ to denote the Banach space of equivalence classes, as well as the vector space of all functions belonging to them, and make the distinction clear when necessary.

If $f(x) \in L_1(-\infty < x < \infty)$, and α any real number, we *define*, for $-\infty < \alpha < \infty$,

$$(1.1) \qquad \hat{f}(\alpha) = \int_{-\infty}^{\infty} f(x) e^{i\alpha x} dx,$$

and say that $\hat{f}$ is the *Fourier transform* of $f \in L_1(-\infty,\infty)$. We shall also use the notation

(1.2) $$F[f] = \frac{1}{\sqrt{(2\pi)}}\hat{f}, \quad \text{or } F[f](\alpha) = \frac{1}{\sqrt{(2\pi)}}\hat{f}(\alpha).$$

(Cf. (8.15) of Ch.I, and §2 of Ch.II.)

In the special case when f is *even*, $f(-x) = f(x)$ for all real values of x, (1.1) takes the form

(1.3) $$\hat{f}(\alpha) = 2\int_0^\infty f(x)\cos\alpha x\,dx.$$

If f is *odd*, $f(-x) = -f(x)$ for all real values of x, (1.1) takes the form

(1.4) $$-i\,\hat{f}(\alpha) = 2\int_0^\infty f(x)\sin\alpha x\,dx.$$

If f is defined in $(0,\infty)$ only, and $f \in L_1(0,\infty)$, then the integrals on the right-hand side of (1.3) and (1.4) define respectively the *cosine transform*, and the *sine transform*, of f.

We list below some basic properties of Fourier transforms of functions in $L_1(-\infty,\infty)$.

(1.5) If $f(x) \in L_1(-\infty < x < \infty)$, then $\hat{f}$ is *bounded* on $(-\infty,\infty)$, since for all real α, we have

$$|\hat{f}(\alpha)| \leq \int_{-\infty}^\infty |f(x)|dx = \| f \|_1 < \infty,$$

where $\| f \|_1$ denotes the L_1-norm of f, so that

$$\sup_{-\infty<\alpha<\infty} |\hat{f}(\alpha)| \leq \| f \|_1 < \infty.$$

(1.6) If $f(x) \in L_1(-\infty < x < \infty)$, then $\hat{f}(\alpha)$ is continuous for $-\infty < \alpha < \infty$. For if h is a real number, $h \neq 0$, then

$$|\hat{f}(\alpha+h) - \hat{f}(\alpha)| = |\int_{-\infty}^\infty f(x)e^{i\alpha x}(e^{ihx}-1)dx|$$

$$\leq \int_{-\infty}^{\infty} |f(x)||e^{ihx}-1|dx,$$

where

$$|f(x)||e^{ihx}-1| \leq 2|f(x)| \in L_1(-\infty,\infty),$$

and

$$|f(x)||e^{ihx}-1| \to 0, \quad \text{as } h \to 0,$$

for almost all $x \in (-\infty,\infty)$. It follows from Lebesgue's theorem on dominated convergence that

$$\int_{-\infty}^{\infty} |f(x)||e^{ihx}-1|dx \to 0, \quad \text{as } h \to 0,$$

and hence $\hat{f}$ is continuous at the point α, where $-\infty < \alpha < \infty$.

(1.7) The operator $f \to \hat{f}$ is linear in the sense that

$$[c_1 f_1 + c_2 f_2]^{\wedge} = c_1 \hat{f}_1 + c_2 \hat{f}_2 \quad ,$$

where c_1, c_2 are complex constants, and f_1, $f_2 \in L_1(-\infty,\infty)$.

(1.8) Let h be a fixed real number, and $f(x) \in L_1(-\infty < x < \infty)$. Then the Fourier transform of $f(x+h)$, the *translation* of $f(x)$ by h, equals $\hat{f}(\alpha)e^{-i\alpha h}$, since

$$\int_{-\infty}^{\infty} f(x+h)e^{i\alpha x}dx = \int_{-\infty}^{\infty} f(x)e^{i\alpha(x-h)}dx = \hat{f}(\alpha)e^{-i\alpha h} .$$

(1.9) Let t be a fixed real number, and $f(x) \in L_1(-\infty < x < \infty)$. Then the Fourier transform of $f(x)e^{itx}$ equals $\hat{f}(\alpha+t)$, since

$$\int_{-\infty}^{\infty} f(x)e^{itx}e^{i\alpha x}dx = \int_{-\infty}^{\infty} f(x)e^{i(t+\alpha)x}dx = \hat{f}(\alpha+t) \quad ,$$

the last being a translation of $\hat{f}(\alpha)$ by t. It follows that *the translation of a Fourier transform is again a Fourier transform.*

(1.10) Let λ be a fixed real number, $\lambda \neq 0$, and $f \in L_1(-\infty,\infty)$. Then the Fourier transform of $f(\lambda x)$ equals

$$\frac{1}{|\lambda|} \hat{f}\left(\frac{\alpha}{\lambda}\right) \quad ,$$

since

$$\int_{-\infty}^{\infty} f(\lambda x)e^{i\alpha x}dx = \frac{1}{|\lambda|}\int_{-\infty}^{\infty} f(y)e^{i(\alpha y)/\lambda}dy.$$

It follows that $\hat{f}$ *is an odd or even function, according as* f *is odd or even.*

(1.11) If $\overline{f}$ denotes the complex conjugate of f, and $f \in L_1(-\infty,\infty)$, then the Fourier transform of $\overline{f}(x)$ equals

$$\overline{\hat{f}(-\alpha)} \quad ,$$

since the complex conjugate of $\int_{-\infty}^{\infty} \overline{f}(x)e^{i\alpha x}dx$ equals $\hat{f}(-\alpha)$.

(1.12) If $f \in L_1(-\infty,\infty)$, and $f_n \in L_1(-\infty,\infty)$ for $n = 1,2,\ldots$, and $\| f_n - f \|_1 \to 0$, as $n \to \infty$, then we have

$$\lim_{n\to\infty} \hat{f}_n(\alpha) = \hat{f}(\alpha) \quad ,$$

uniformly for $-\infty < \alpha < \infty$, since by (1.5) and (1.7), we have

$$\sup_{-\infty<\alpha<\infty} |\hat{f}_n(\alpha) - \hat{f}(\alpha)| \leq \| f_n - f \|_1 \quad .$$

(1.13) If $f_1, f_2 \in L_1(-\infty,\infty)$, then we have the *composition rule*

$$\int_{-\infty}^{\infty} \hat{f}_1(y)f_2(y)dy = \int_{-\infty}^{\infty} f_1(y)\hat{f}_2(y)dy \quad ,$$

since

$$\int_{-\infty}^{\infty} \hat{f}_1(y)f_2(y)dy = \int_{-\infty}^{\infty} f_2(y)\left(\int_{-\infty}^{\infty} f_1(x)e^{iyx}dx\right)dy$$

$$= \int_{-\infty}^{\infty}\int_{-\infty}^{\infty} f_2(y)f_1(x)e^{iyx}dx\,dy \quad ,$$

by Fubini's theorem, and

$$\int_{-\infty}^{\infty}\int_{-\infty}^{\infty} |f_2(y)||f_1(x)|dx\,dy = \int_{-\infty}^{\infty} |f_1(x)|dx \int_{-\infty}^{\infty} |f_2(y)|dy < \infty.$$

The integral

$$\int_{-\infty}^{\infty}\int_{-\infty}^{\infty} f_2(y)f_1(x)e^{iyx}dx\,dy$$

is symmetric in f_1 and f_2, so that we can interchange f_2 and f_1, and assertion (1.13) follows.

A fundamental property of Fourier transforms of functions in $L_1(-\infty,\infty)$ is contained in the following

Theorem 1 (Riemann-Lebesgue). If $f(x) \in L_1(-\infty < x < \infty)$, *and* $\hat{f}$ *denotes the Fourier transform of* f, *then*

$$\hat{f}(\alpha) \to 0, \quad \text{as } |\alpha| \to \infty.$$

Proof. Consider the special function χ defined by

$$\chi(x) = \begin{cases} 1, & \text{for } -\infty < a \leq x \leq b < \infty, \\ 0, & \text{for } x < a, \text{ and } x > b, \end{cases}$$

referred to as the *characteristic function*, or *indicator function*, of the finite interval [a,b]. Its Fourier transform is

$$\hat{\chi}(\alpha) = \int_a^b f(x) e^{i\alpha x} dx = \frac{1}{i\alpha}\left(e^{i\alpha b} - e^{i\alpha a}\right),$$

for α real, $\alpha \neq 0$, so that

$$|\hat{\chi}(\alpha)| \leq \frac{2}{|\alpha|} \to 0, \quad \text{as } |\alpha| \to \infty.$$

By linearity, this property holds also for any *step-function* (such a function being a finite linear combination, with complex coefficients, of characteristic (or indicator) functions of finite intervals). And step-functions form a *dense* subset of $L_1(-\infty,\infty)$. That is to say, given $\varepsilon > 0$, and $f \in L_1(-\infty,\infty)$, there exists a step-function f_ε, such that $\| f - f_\varepsilon \|_1 < \varepsilon$. Since

$$\hat{f}(\alpha) = \hat{f}_\varepsilon(\alpha) + (\hat{f}(\alpha) - \hat{f}_\varepsilon(\alpha)),$$

we have

$$|\hat{f}(\alpha)| \leq |\hat{f}_\varepsilon(\alpha)| + |\hat{f}(\alpha) - \hat{f}_\varepsilon(\alpha)| < |\hat{f}_\varepsilon(\alpha)| + \varepsilon,$$

because of property (1.5) and the choice of f_ε. Hence

$$\limsup_{|\alpha|\to\infty} |\hat{f}(\alpha)| \le \limsup_{|\alpha|\to\infty} |\hat{f}_\varepsilon(\alpha)| + \varepsilon = \varepsilon,$$

since we have just seen that the theorem holds for the step-function f_ε.

<u>*Corollary.*</u> *If* $f \in L_1(-\infty,\infty)$, *then we have*

$$\int_{-\infty}^{\infty} f(t)\cos(\alpha t)\,dt \to 0, \quad \textit{as } |\alpha| \to \infty,$$

and

$$\int_{-\infty}^{\infty} f(t)\sin(\alpha t)\,dt \to 0, \quad \textit{as } |\alpha| \to \infty.$$

<u>Remarks</u>. 1. The Riemann-Lebesgue theorem is related to a property of the L_p-*modulus of continuity* τ_f of the function $f \in L_p(-\infty,\infty)$, $1 \le p < \infty$, which is defined by

$$(1.14) \qquad \tau_f(y) = \left(\int_{-\infty}^{\infty} |f(x) - f(x+y)|^p dx\right)^{1/p}, \quad -\infty < y < \infty.$$

Clearly we have

$$(1.15) \qquad \tau_f(y) = \tau_f(-y), \ \tau_f(0) = 0, \quad 0 \le \tau_f(y) \le 2\|f\|_p,$$

and

$$(1.16) \qquad \tau_f(y_1 + y_2) \le \tau_f(y_1) + \tau_f(y_2),$$

since

$$\left(\int_{-\infty}^{\infty} |f(x) - f(x+y_1+y_2)|^p dx\right)^{1/p}$$

$$= \left(\int_{-\infty}^{\infty} |f(x) - f(x+y_1) + f(x+y_1) - f(x+y_1+y_2)|^p dx\right)^{1/p}$$

$$\le \tau_f(y_1) + \tau_f(y_2) \quad .$$

(1.17) If $f \in L_p(-\infty,\infty)$, $1 \le p < \infty$, then $\tau_f(h)$ is continuous in h; in particular, $\tau_f(h) \to 0$, as $h \to 0$.

To prove this we note that given $\varepsilon > 0$, there exists a continuous function $\varphi(x)$ which vanishes outside a finite interval, such that

$$\left(\int_{-\infty}^{\infty} |f(x) - \varphi(x)|^p dx\right)^{1/p} < \varepsilon.$$

If we set

$$\Phi(h) = \left(\int_{-\infty}^{\infty} |\varphi(x+h) - \varphi(x)|^p dx\right)^{1/p},$$

then $\Phi(h)$ is continuous in h, and

$$\tau_f(h) \leq \left(\int_{-\infty}^{\infty} |f(x+h) - \varphi(x+h)|^p dx\right)^{1/p}$$

$$+ \left(\int_{-\infty}^{\infty} |\varphi(x+h) - \varphi(x)|^p dx\right)^{1/p} + \left(\int_{-\infty}^{\infty} |\varphi(x) - f(x)|^p dx\right)^{1/p},$$

so that

$$\tau_f(h) - \Phi(h) < 2\varepsilon,$$

and similarly also

$$\Phi(h) - \tau_f(h) < 2\varepsilon,$$

so that $\Phi \to \tau_f$ uniformly in h as $\varepsilon \downarrow 0$. Hence $\tau_f(h)$ is continuous in h, and tends to zero with h.

Reverting to the proof of Theorem 1, if $f \in L_1(-\infty,\infty)$, then

$$\hat{f}(\alpha) = \int_{-\infty}^{\infty} f(x)e^{i\alpha x}dx, \quad -\hat{f}(\alpha) = \int_{-\infty}^{\infty} f(x)e^{i(x+\pi/\alpha)\alpha}dx$$

$$= \int_{-\infty}^{\infty} f(y-\tfrac{\pi}{\alpha})e^{i\alpha y}dy,$$

so that

$$|2\hat{f}(\alpha)| \leq \int_{-\infty}^{\infty} |f(y) - f(y-\tfrac{\pi}{\alpha})|dy,$$

and, because of (1.17), $\hat{f}(\alpha) \to 0$, as $|\alpha| \to \infty$.

2. If $f \in L_1(-\infty,\infty)$, the Fourier transform $\hat{f}(\alpha)$ is a continuous function of α, $-\infty < \alpha < \infty$, which *vanishes at* infinity; that is to say, $\lim_{|\alpha|\to\infty} \hat{f}(\alpha) = 0$. Not every continuous function that vanishes at in-

finity is necessarily the Fourier transform of a function in $L_1(-\infty,\infty)$.

To construct an example, let $f \in L_1(-\infty,\infty)$, and let f be *odd*, that is to say $f(-x) = -f(x)$, $-\infty < x < \infty$. Then obviously we have

$$\hat{f}(\alpha) = 2i \int_0^\infty f(x) \sin \alpha x \, dx .$$

Let e denote the exponential, and $R > e$. Then we have

$$\int_e^R \frac{\hat{f}(\alpha)}{\alpha} d\alpha = 2i \int_e^R \frac{d\alpha}{\alpha} \int_0^\infty f(x) \sin \alpha x \, dx$$

$$= 2i \int_0^\infty f(x) \left(\int_e^R \frac{\sin \alpha x}{\alpha} d\alpha \right) dx$$

$$= 2i \int_0^\infty f(x) \left(\int_{ex}^{Rx} \frac{\sin \alpha}{\alpha} d\alpha \right) dx,$$

the change in the order of integration being permitted by Fubini's theorem, because of the assumption $f \in L_1(-\infty,\infty)$. Now

$$\left| \int_a^b \frac{\sin \alpha}{\alpha} d\alpha \right| < M_0 < \infty$$

for all real a,b, where M_0 is independent of a and b, and hence

$$\left| \int_e^R \frac{\hat{f}(\alpha)}{\alpha} d\alpha \right| < M < \infty. \tag{1.18}$$

If we define the function g by

$$g(\alpha) = \begin{cases} \alpha/e, & 0 \le \alpha \le e, \\ 1/\log \alpha, & \alpha > e; \end{cases} \qquad g(\alpha) = -g(-\alpha), \ \alpha \ge 0;$$

then g is an odd, continuous function, which vanishes at infinity, for which

$$\int_e^R \frac{g(\alpha)}{\alpha} d\alpha = \int_e^R \frac{d\alpha}{\alpha \log \alpha} = \log\log R \to \infty, \text{ as } R \to \infty.$$

Clearly g cannot be the Fourier transform of a function in $L_1(-\infty,\infty)$, since it cannot take the place of $\hat{f}$ in (1.18).

Examples

1. If

$$f(x) = \begin{cases} 1, & \text{for } |x| \le 1, \\ 0, & \text{for } |x| > 1, \end{cases}$$

then

$$\hat{f}(\alpha) = \int_{-1}^{1} e^{i\alpha x}dx = 2\int_{0}^{1} \cos \alpha x \, dx = \left[2 \frac{\sin \alpha x}{\alpha}\right]_{x=0}^{x=1} = 2\left(\frac{\sin\alpha}{\alpha}\right).$$

Note that here $\hat{f}(\alpha) \notin L_1(-\infty,\infty)$.

Similarly if

$$f(x) = \begin{cases} 1, & \text{for } |x-a| \le R, \quad -\infty < a < \infty,\ R > 0, \\ 0, & \text{for } |x-a| > R, \end{cases}$$

then $\hat{f}(\alpha) = 2\, e^{i\alpha a} \dfrac{\sin \alpha R}{\alpha}$.

2. If

$$f(x) = \begin{cases} 1-|x|, & \text{for } |x| \le 1, \\ 0, & \text{for } |x| > 1, \end{cases}$$

then

$$\hat{f}(\alpha) = \left(\frac{\sin (\alpha/2)}{\alpha/2}\right)^2,$$

since

$$\hat{f}(\alpha) = 2\int_{0}^{1} (1-x)\cos \alpha x \, dx = 2\int_{0}^{1} (1-x) \frac{d}{dx}\left(\frac{\sin \alpha x}{\alpha}\right)dx$$

$$= 2\int_{0}^{1} \frac{\sin \alpha x}{\alpha}\, dx = \int_{0}^{1} \frac{d}{dx}\left(\frac{\sin^2(\alpha x/2)}{(\alpha/2)^2}\right)dx = \frac{\sin^2(\alpha/2)}{(\alpha/2)^2}.$$

Note that here $\hat{f}(\alpha) \in L_1(-\infty,\infty)$.

Similarly if $a > 0$, and $f(x) = 1 - \frac{|x|}{a}$, for $|x| \le a$, and $f(x) = 0$, for $|x| > a$, then $\hat{f}(\alpha) = \dfrac{\sin^2(\alpha a/2)}{a(\frac{\alpha}{2})^2}$.

3. If $f(x) = e^{-e^x} e^x$, then $\hat{f}(\alpha) = \Gamma(1+i\alpha) \neq 0$, where Γ stands for Euler's gamma-function, since

$$\hat{f}(\alpha) = \int_{-\infty}^{\infty} e^{-e^x} e^x e^{i\alpha x} dx = \int_0^{\infty} e^{-y} y^{i\alpha} dy = \Gamma(1+i\alpha).$$

Similarly if $f(x) = e^{-e^{-x}} e^{-x}$, then $\hat{f}(\alpha) = \Gamma(1-i\alpha)$.

4. If $f(x) = e^{-|x|}$, then $\hat{f}(\alpha) = \frac{2}{1+\alpha^2}$. For

$$\int_0^{\infty} e^{-x+i\alpha x} dx = \frac{1}{1-i\alpha}; \quad \int_{-\infty}^{0} e^{-|x|+i\alpha x} dx = \frac{1}{1+i\alpha} ;$$

and the result follows from adding these two integrals. Here again we have $\hat{f}(\alpha) \in L_1(-\infty,\infty)$.

If $a > 0$, and $f(x) = e^{-a|x|}$, then $\hat{f}(\alpha) = \frac{2a}{a^2+\alpha^2}$.

5. Let ζ be a complex number, $\zeta = \xi + i\eta$, with $\eta \neq 0$. If $\eta > 0$, let

$$f(x) = \begin{cases} 0, & \text{for } x > 0, \\ e^{-ix\zeta}, & \text{for } x < 0; \end{cases}$$

and if $\eta < 0$, let

$$f(x) = \begin{cases} 0, & \text{for } x < 0, \\ -e^{-ix\zeta}, & \text{for } x > 0. \end{cases}$$

Then

$$\hat{f}(\alpha) = \frac{1}{i(\alpha-\zeta)} ,$$

for, if $\eta > 0$,

$$\int_{-\infty}^{\infty} f(x) e^{i\alpha x} dx = \int_{-\infty}^{0} e^{-ix\zeta+i\alpha x} dx = \int_0^{\infty} e^{-x(\eta-i\xi+i\alpha)} dx = \frac{1}{i(\alpha-\zeta)} ,$$

and, if $\eta < 0$,

$$\int_{-\infty}^{\infty} f(x) e^{i\alpha x} dx = -\int_0^{\infty} e^{-ix\zeta+i\alpha x} dx = -\int_0^{\infty} e^{-x(-\eta+i\xi-i\alpha)} dx = \frac{1}{i(\alpha-\zeta)} .$$

6. If $f(x) = e^{-x^2}$, then $\hat{f}(\alpha) = \sqrt{\pi}\, e^{-\alpha^2/4}$.

To prove this, we note, first of all, that

$$\int_{-\infty}^{\infty} e^{-x^2}dx \int_{-\infty}^{\infty} e^{-y^2}dy = \int_{-\infty}^{\infty}\int_{-\infty}^{\infty} e^{-(x^2+y^2)}dx\,dy = \int_0^{2\pi} d\theta \int_0^{\infty} e^{-r^2} r\,dr$$

$$= 2\pi \int_0^{\infty} \frac{1}{2} e^{-r^2} d(r^2) = \pi ,$$

so that

$$\int_{-\infty}^{\infty} e^{-x^2} dx = \sqrt{\pi}.$$

Next we note that the derivative of $\hat{f}(\alpha)$ is given by

$$(\hat{f}(\alpha))' \equiv \frac{d}{d\alpha}(\hat{f}(\alpha)) = i \int_{-\infty}^{\infty} x\, e^{-x^2} e^{i\alpha x}dx = -\frac{i}{2} \int_{-\infty}^{\infty} \frac{d}{dx}(e^{-x^2})e^{i\alpha x}dx$$

$$= (\tfrac{i}{2}) \int_{-\infty}^{\infty} e^{-x^2}\, i\alpha\, e^{i\alpha x}dx = (-\tfrac{\alpha}{2}) \int_{-\infty}^{\infty} e^{-x^2} e^{i\alpha x}dx ,$$

so that

$$(\hat{f}(\alpha))' = (-\tfrac{\alpha}{2})\ \hat{f}(\alpha), \text{ or } \frac{(\hat{f}(\alpha))'}{\hat{f}(\alpha)} = -\alpha/2 ,$$

and hence

$$\log \hat{f}(\alpha) = -(\alpha^2/4) + c_1, \text{ or } \hat{f}(\alpha) = c\, e^{-\alpha^2/4} ,$$

where c_1, c, are constants. On setting $\alpha = 0$, we see that $\hat{f}(0) = c = \sqrt{\pi}$, because of the evaluation of $\int_{-\infty}^{\infty} e^{-x^2} dx$ first made.

It follows that if $f(x) = e^{-\frac{1}{2}x^2}$, then $\hat{f}(\alpha) = \sqrt{2\pi}\, e^{-\frac{1}{2}\alpha^2}$.

7\. Let $\varphi_n(x) = (-1)^n e^{\frac{1}{2}x^2}\left(\frac{d}{dx}\right)^n e^{-x^2}$, where n is a positive integer. Then

$$\hat{\varphi}_n(\alpha) = (-i)^n \sqrt{2\pi}\, e^{\frac{1}{2}\alpha^2}\left(\frac{d}{d\alpha}\right)^n e^{-\alpha^2} = i^n \sqrt{2\pi}\, \varphi_n(\alpha),$$

for

$$\int_{-\infty}^{\infty}\varphi_n(x)e^{i\alpha x}dx = (-1)^n \int_{-\infty}^{\infty} e^{\frac{1}{2}x^2}\left(\frac{d}{dx}\right)^n e^{-x^2}e^{ix\alpha}dx$$

$$= \int_{-\infty}^{\infty} e^{-x^2}\left(\frac{d}{dx}\right)^n\left(e^{\frac{1}{2}x^2+ix\alpha}\right)dx \text{ (by partial integration)}$$

$$= e^{\frac{1}{2}\alpha^2}\int_{-\infty}^{\infty} e^{-x^2}\left(\frac{d}{dx}\right)^n\left(e^{\frac{1}{2}(x+i\alpha)^2}\right)dx$$

$$= (-i)^n e^{\frac{1}{2}\alpha^2}\int_{-\infty}^{\infty} e^{-x^2}\left(\frac{d}{d\alpha}\right)^n\left(e^{\frac{1}{2}(x+i\alpha)^2}\right)dx$$

$$= (-i)^n e^{\frac{1}{2}\alpha^2}\left(\frac{d}{d\alpha}\right)^n\left(\int_{-\infty}^{\infty} e^{-\frac{1}{2}x^2+i\alpha x-\alpha^2/2}dx\right)$$

$$= (-i)^n e^{\frac{1}{2}\alpha^2}\left(\frac{d}{d\alpha}\right)^n\left(\sqrt{(2\pi)}\; e^{-\alpha^2}\right) = i^n\sqrt{(2\pi)}\;\varphi_n(\alpha) \;.$$

The functions $(\varphi_n(x))$ are known as the *Hermite functions*.

8. If
$$f(x) = \begin{cases} 0, & \text{for } |x| \geq 1, \\ (1-x^2)^{\nu-\frac{1}{2}}, & \text{for } 0 < |x| < 1,\ \nu > -\frac{1}{2}\,, \end{cases}$$

then $f \in L_1(-\infty,\infty)$, and

$$\hat{f}(\alpha) = 2\int_0^1 (1-x^2)^{\nu-\frac{1}{2}}\cos(\alpha x)\,dx$$

$$= 2\sum_{n=0}^{\infty}\frac{(-1)^n\alpha^{2n}}{(2n)!}\int_0^1 (1-x^2)^{\nu-\frac{1}{2}}x^{2n}dx$$

$$= \sum_{n=0}^{\infty}\frac{(-1)^n\alpha^{2n}}{(2n)!}\;\frac{\Gamma(\nu+\frac{1}{2})\Gamma(n+\frac{1}{2})}{\Gamma(\nu+n+1)}$$

$$= \sqrt{\pi}\;\Gamma(\nu+\tfrac{1}{2})\sum_{n=0}^{\infty}\frac{(-1)^n\alpha^{2n}}{2^{2n}\;n!\;\Gamma(\nu+n+1)}\;,$$

since

$$\Gamma(n+\tfrac{1}{2}) = (n-\tfrac{1}{2})(n-\tfrac{3}{2})\;\ldots\;\tfrac{1}{2}\;\Gamma(\tfrac{1}{2}) = \frac{(2n-1)(2n-3)\;\ldots\;5.3.1}{2^n}\sqrt{\pi}$$

$$= \frac{(2n)!\sqrt{\pi}}{2^{2n}(n!)}\;.$$

The *Bessel function* J_ν of order ν is *defined* by

$$J_\nu(x) = \sum_{n=0}^{\infty} \frac{(-1)^n (\frac{1}{2}x)^{\nu+2n}}{n!\,\Gamma(\nu+n+1)}, \text{ for } \nu > -1.$$

Hence

$$\hat{f}(\alpha) = \frac{J_\nu(\alpha)}{\alpha^\nu}\, 2^\nu \Gamma(\nu+\tfrac{1}{2})\sqrt{\pi} .$$

Incidentally we obtain integral representations for $J_\nu(x)$, namely

$$J_\nu(\alpha) = \frac{(\alpha/2)^\nu}{\Gamma(\nu+\frac{1}{2})\sqrt{\pi}} \int_{-1}^{1} (1-t^2)^{\nu-\frac{1}{2}} e^{it\alpha} dt, \quad \nu > -\tfrac{1}{2},$$

and, on setting $t = \cos\theta$,

$$J_\nu(\alpha) = \frac{(\alpha/2)^\nu}{\Gamma(\nu+\frac{1}{2})\sqrt{\pi}} \int_{0}^{\pi} \sin^{2\nu}\theta\; e^{i\alpha\cos\theta} d\theta, \quad \nu > -\tfrac{1}{2}.$$

9. If $f(x) = \dfrac{1}{\cosh \pi x}$, then $f(x) \in L_1(-\infty < x < \infty)$, and $\hat{f}(\alpha) = \dfrac{1}{\cosh(\alpha/2)}$.

If we apply Cauchy's theorem to the integral $\int_C \dfrac{e^{ix\alpha}}{\cosh \pi x} dx$ taken along the contour C, which is a rectangle with corners at $-R$, $+R$, $R+i$, and $-R+i$ (where $R > 0$), and note that the residue at the pole $x = i/2$ equals $e^{-\alpha/2}/(\pi i)$, and then let $R \to \infty$, we get

$$\frac{1}{2\pi i}\left[\int_{-\infty}^{\infty} \frac{e^{ix\alpha} dx}{\cosh \pi x} - \int_{-\infty}^{\infty} \frac{e^{i(x+i)\alpha} dx}{\cosh \pi(x+i)}\right] = \frac{e^{-\alpha/2}}{\pi i},$$

so that

$$\hat{f}(\alpha)\left[1+e^{-\alpha}\right] = 2\, e^{-\alpha/2}, \text{ or } \hat{f}(\alpha) = \left[\cosh(\alpha/2)\right]^{-1}.$$

10. <u>*An integral of Ramanujan.*</u>

If $f(x) = \dfrac{e^{-i\pi x^2}}{\cosh \pi x}$, then $f(x) \in L_1(-\infty < x < \infty)$, and

$$\varphi(\alpha) \equiv \int_{-\infty}^{\infty} \frac{e^{-i\pi x^2}\, e^{-ix\alpha}}{\cosh \pi x} dx = \frac{e^{\frac{i\pi}{4}} - ie^{\frac{i\alpha^2}{4\pi}}}{\cosh(\alpha/2)} .$$

<u>Proof</u>. It is immediate that

(i) $$\varphi(\alpha+i\pi) + \varphi(\alpha-i\pi) = 2\int_{-\infty}^{\infty} e^{-i\pi x^2 - ix\alpha}dx .$$

Since $e^{-i\pi x^2} \notin L_1(-\infty < x < \infty)$, the integral on the right-hand side is defined as a Cauchy principal value (cf. §4). To evaluate it one can use Cauchy's theorem, or use Example 6, which gives

(ii) $$\int_{-\infty}^{\infty} e^{-\lambda t^2 + it\alpha}dt = \frac{\sqrt{\pi}}{\sqrt{\lambda}} e^{-\alpha^2/(4\lambda)}, \quad \text{for } \lambda > 0.$$

By analytic continuation, this holds good also for *complex* λ, with $\operatorname{Re}\lambda \geq 0$, provided that $\lambda \neq 0$. Thus, for $\lambda = \pm\pi i$, we get:

(iii) $$\int_{-\infty}^{\infty} e^{-i\pi t^2 - it\alpha}dt = e^{\frac{i\alpha^2}{4\pi} - \frac{\pi i}{4}} ; \int_{-\infty}^{\infty} e^{i\pi t^2 - it\alpha}dt = e^{\frac{-i\alpha^2}{4\pi} + \frac{\pi i}{4}} .$$

On using the first formula in (i), we obtain

(iv) $$\varphi(\alpha+i\pi) + \varphi(\alpha-i\pi) = 2e^{i\alpha^2/(4\pi) - \frac{\pi i}{4}} .$$

We shall see, on the other hand, that

(v) $$e^{\alpha/2}\varphi(\alpha+i\pi) + e^{-\alpha/2}\varphi(\alpha-i\pi) = 2e^{\frac{-i\pi}{4}} .$$

To prove this, we note that (ii) implies that

(vi) $$\int_{-\infty}^{\infty} e^{-\lambda(t+x)^2 + it\alpha}dt = \frac{\sqrt{\pi}}{\sqrt{\lambda}} e^{-\alpha^2/(4\lambda)} e^{-ix\alpha}, \quad \text{for } \operatorname{Re}\lambda > 0.$$

Let $\varepsilon > 0$, and $\lambda = 1/[4(\varepsilon+i\pi)]$. Using the composition rule (1.13) together with the Fourier transforms given in Example 9, and (vi), we get

$$2\pi^{\frac{1}{2}} (\varepsilon+\pi i)^{\frac{1}{2}} \int_{-\infty}^{\infty} \frac{e^{-(\varepsilon+i\pi)x^2 - ix\alpha}}{\cosh \pi x} dx = \int_{-\infty}^{\infty} \frac{e^{-\lambda(x+\alpha)^2}}{\cosh(x/2)} dx,$$

$$\lambda = 1/[4(\varepsilon+\pi i)], \ \operatorname{Re}\lambda > 0.$$

Lebesgue's theorem on dominated convergence permits us to let $\varepsilon \downarrow 0$ in this relation, and we obtain

$$\varphi(\alpha) = e^{\frac{-\pi i}{4}} \frac{1}{2\pi} \int_{-\infty}^{\infty} \frac{e^{i(x+\alpha)^2/(4\pi)}}{\cosh(x/2)} dx$$

$$= e^{\frac{-\pi i}{4}} \int_{-\infty}^{\infty} \frac{e^{i(2\pi y+\alpha)^2/(4\pi)}}{\cosh(\pi y)} dy, \qquad x = 2\pi y,$$

$$= e^{\frac{-\pi i}{4}} \int_{-\infty}^{\infty} \frac{e^{i(-2\pi y+\alpha)^2/(4\pi)}}{\cosh \pi y} dy$$

$$= e^{\frac{i\alpha^2}{4\pi} - \frac{\pi i}{4}} \int_{-\infty}^{\infty} \frac{e^{i\pi y^2 - i\alpha y}}{\cosh \pi y} dy .$$

This leads, as before, to the relation

$$e^{-i(\alpha+i\pi)^2/(4\pi)}\varphi(\alpha+i\pi) + e^{-i(\alpha-i\pi)^2/(4\pi)}\varphi(\alpha-i\pi) = 2e^{\frac{-\pi i}{4}} \int_{-\infty}^{\infty} e^{i\pi y^2 - i\alpha y} dy$$

$$= 2e^{-i\alpha^2/(4\pi)} ,$$

if we use the second formula in (iii), and this leads to (v).

On multiplying (iv) by $e^{-\alpha/2}$, and substituting from (v), we obtain

$$\varphi(\alpha+i\pi)\left(e^{\alpha/2} - e^{-\alpha/2}\right) = 2e^{\frac{-\pi i}{4}} \left(1 - e^{\frac{i\alpha^2}{4\pi} - \frac{\alpha}{2}}\right) ,$$

and on replacing α by $\alpha-i\pi$, we get

$$\varphi(\alpha) = 2e^{\frac{-\pi i}{4}} \frac{\left[1 - e^{\frac{i(\alpha-i\pi)^2}{4\pi} - \frac{(\alpha-i\pi)}{2}}\right]}{e^{(\alpha-i\pi)/2} - e^{-(\alpha-i\pi)/2}} = \frac{e^{\frac{\pi i}{4}} - i\, e^{\frac{i\alpha^2}{4\pi}}}{\cosh(\alpha/2)} ,$$

as claimed.

On equating the real and imaginary parts of both sides, and on setting $\alpha = 2\pi t$, we get Ramanujan's formulae:

$$\int_0^\infty \frac{\cos 2\pi tx}{\cosh \pi x} \cos \pi x^2 dx = \frac{1 + \sqrt{2} \sin \pi t^2}{2\sqrt{2} \cosh \pi t} ,$$

$$\int_0^\infty \frac{\cos 2\pi tx}{\cosh \pi x} \sin \pi x^2 dx = \frac{-1 + \sqrt{2}\cos \pi t^2}{2\sqrt{2}\cosh \pi t} .$$

§2. The L_1-algebra

The Banach space $L_1(-\infty,\infty)$ can be made into a *Banach algebra* by the introduction of an operation of 'multiplication', which is defined by the *convolution* of any two functions. To make this possible we need the following

(2.1) Lemma. *If* $f,g \in L_1(-\infty,\infty)$, *then the integral*

$$\int_{-\infty}^{\infty} f(x-y)g(y)dy \qquad \left(= \int_{-\infty}^{\infty} f(y)g(x-y)dy\right)$$

exists for almost all $x \in (-\infty,\infty)$, *and is an integrable function of* x, $-\infty < x < \infty$.

Proof. The function $f(x-y)g(y)$ is a measurable function in (x,y), and the double integral

$$\int_{-\infty}^{\infty}\int_{-\infty}^{\infty} |f(x-y)g(y)|dx\,dy \leq \infty.$$

By Fubini's theorem, this integral equals the repeated integral

$$\int_{-\infty}^{\infty} |g(y)|\left(\int_{-\infty}^{\infty} |f(x-y)|dx\right)dy = \int_{-\infty}^{\infty} |g(y)|\left(\int_{-\infty}^{\infty} |f(x)|dx\right)dy < \infty,$$

and hence we have also

$$\int_{-\infty}^{\infty}\left(\int_{-\infty}^{\infty} |f(x-y)g(y)|dy\right)dx < \infty,$$

so that $|f(x-y)g(y)|$ is integrable as a function of y for almost all $x \in (-\infty,\infty)$, and

$$\int_{-\infty}^{\infty} |f(x-y)g(y)|dy$$

is integrable as a function of x, and therefore also $\int_{-\infty}^{\infty} f(x-y)g(y)dy$.

<u>Definition</u>. Let $f,g \in L_1(-\infty,\infty)$, and let

$$h(x) = \int_{-\infty}^{\infty} f(x-y)g(y)dy,$$

when the integral exists. Then the function h is defined to be the convolution of f and g, and is denoted by $h = f*g$.

By Lemma 2.1 we note that $h \in L_1(-\infty,\infty)$. We also note that the convolution is *associative*:

$$(f*g)*h = f*(g*h),$$

where $f,g,h \in L_1(-\infty,\infty)$. It is also commutative, that is to say $f*g = g*f$ for $f,g \in L_1(-\infty,\infty)$ as can be seen by a change of variable in the defining integral.

<u>*Theorem 2*</u>. *If* $f,g \in L_1(-\infty,\infty)$, *and* $h = f*g$, *then* $h \in L_1(-\infty,\infty)$, *and*

$$\hat{h} = \hat{f}\cdot\hat{g}, \tag{2.2}$$

where the dot denotes pointwise multiplication, and

$$\| h \|_1 = \| f*g \|_1 \leq \| f \|_1 \cdot \| g \|_1. \tag{2.3}$$

<u>Proof</u>. Lemma 2.1 shows that $h \in L_1(-\infty,\infty)$. On using the definition of h and of $\hat{h}$, together with Fubini's theorem, we see that

$$\hat{h}(\alpha) = \int_{-\infty}^{\infty} h(x)e^{i\alpha x}dx = \int_{-\infty}^{\infty} e^{i\alpha x}dx \int_{-\infty}^{\infty} f(x-y)g(y)dy$$

$$= \int_{-\infty}^{\infty} g(y)dy \int_{-\infty}^{\infty} f(x-y)e^{i\alpha x}dx = \int_{-\infty}^{\infty} g(y)dy \int_{-\infty}^{\infty} f(u)e^{i(u+y)\alpha}du$$

$$= \int_{-\infty}^{\infty} e^{i\alpha y}g(y)dy \int_{-\infty}^{\infty} f(u)e^{iu\alpha}du = \hat{g}(\alpha)\cdot\hat{f}(\alpha).$$

On the other hand, we have

$$\| h \|_1 = \int_{-\infty}^{\infty} |h(x)|dx \leq \int_{-\infty}^{\infty} dx \int_{-\infty}^{\infty} |f(x-y)g(y)|dy$$

$$= \int_{-\infty}^{\infty} |g(y)|dy \int_{-\infty}^{\infty} |f(x-y)|dy = \| g \|_1 \cdot \| f \|_1 .$$

Remarks

The properties of convolution just established show that the space $L_1(-\infty,\infty)$ is a commutative Banach algebra under ordinary addition, with convolution as multiplication, and $\| \cdot \|_1$ as *norm*. This Banach algebra is also known as the L_1-*algebra on the real line.*

§3. Properties of differentiability

We have seen that if a function f belongs to the Lebesgue class $L_1(-\infty,\infty)$, then its Fourier transform $\hat{f}$ is a bounded, continuous function on $(-\infty,\infty)$. By imposing a simple additional condition on f, one can secure the differentiability of $\hat{f}$. If, on the other hand, the function $f \in L_1(-\infty,\infty)$ itself is assumed to be differentiable, and the derivative also belongs to $L_1(-\infty,\infty)$, then the Fourier transform of the derivative f' is related, in a simple way, to the transform of f itself.

Theorem 3. A. If $f(x) \in L_1(-\infty < x < \infty)$, *and* $ix\cdot f(x) \in L_1(-\infty < x < \infty)$, *then* $\hat{f}(\alpha)$ *exists, and*

$$(\hat{f})'(\alpha) = \int_{-\infty}^{\infty} [ix\cdot f(x)]e^{i\alpha x}dx .$$

B. If $f(x) \in L_1(-\infty < x < \infty)$, f *is continuously differentiable, and* $f'(x) \in L_1(-\infty < x < \infty)$, *then*

$$\int_{-\infty}^{\infty} [f'(x)]e^{i\alpha x}dx = -i\alpha\cdot\hat{f}(\alpha),$$

so that

$$\hat{f}(\alpha) = o\left(\frac{1}{|\alpha|}\right), \text{ as } |\alpha| \to \infty.$$

Proof. A. Let

$$f_h(x) = f(x)\left(\frac{e^{ihx}-1}{h}\right)$$

for h real, and $h \neq 0$. Then

$$(3.1) \qquad \hat{f}_h(\alpha) = \frac{\hat{f}(\alpha+h) - \hat{f}(\alpha)}{h} .$$

Now $f_h(x) \to ix\cdot f(x)$ pointwise, for almost every x, as $h \to 0$; and

$$|f_h(x)| \leq |f(x)|\cdot\left|\frac{e^{ihx}-1}{h}\right| \leq |x|\cdot|f(x)| \in L_1(-\infty < x < \infty),$$

by hypothesis (on using the first mean-value theorem). Hence

$$|f_h(x) - ix\cdot f(x)| \leq 2|x|\cdot|f(x)| \in L_1(-\infty,\infty).$$

By Lebesgue's theorem on dominated convergence, we conclude that

$$f_h(x) \to ix\cdot f(x), \text{ in the } L_1\text{-norm},$$

as $h \to 0$. By property (1.12) it follows that

$$(3.2) \qquad \hat{f}_h(\alpha) \to \int_{-\infty}^{\infty} [ix\cdot f(x)]e^{i\alpha x}dx,$$

uniformly, as $h \to 0$. The last integral is a bounded, continuous function of α for $-\infty < \alpha < \infty$. From (3.1), however, we see that

$$\lim_{h\to 0} \hat{f}_h(\alpha) = (\hat{f})'(\alpha) ,$$

hence

$$(\hat{f})'(\alpha) = \int_{-\infty}^{\infty} \{ix\cdot f(x)\}e^{i\alpha x}dx.$$

B. Let α be real, $\alpha \neq 0$. For $R > 0$ we have

$$(3.3) \qquad \int_{-R}^{R} f(x)e^{i\alpha x}dx = \left[\frac{e^{i\alpha x}}{i\alpha} f(x)\right]_{x=-R}^{x=R} - \int_{-R}^{R} \frac{e^{i\alpha x}}{i\alpha} f'(x)dx.$$

Now f(x) tends to a *finite* limit as $x \to \pm\infty$. For

$$f(x) - f(0) = \int_0^x f'(t)dt, \textit{ and } f' \in L_1(-\infty,\infty),$$

so that

$$f(\pm\infty) = f(0) + \int_0^{\pm\infty} f'(t)dt.$$

But $f(\pm\infty) = 0$. Hence, on letting $R \to \infty$ in (3.3), we have

$$\int_{-\infty}^{\infty} e^{i\alpha x} f(x)\,dx = -\int_{-\infty}^{\infty} \frac{e^{i\alpha x}}{i\alpha} f'(x)\,dx, \quad \alpha \neq 0,$$

which implies that

$$(3.4) \qquad (-i\alpha)\hat{f}(\alpha) = (f')^{\wedge}(\alpha), \text{for } \alpha \neq 0.$$

This holds also for $\alpha = 0$ by continuity (since the left-hand side is zero, while the right-hand side is

$$\int_{-\infty}^{\infty} f'(x)\,dx = f(\infty) - f(-\infty) = 0).$$

Since

$$|\alpha\hat{f}(\alpha)| \leq \int_{-\infty}^{\infty} |f'(x)|\,dx = \| f' \|_1 < \infty,$$

and $\hat{f}(\alpha) = o(1)$, as $|\alpha| \to \infty$, by Theorem 1, we conclude that $\hat{f}(\alpha) = O(1/|\alpha|)$. Actually $\hat{f}(\alpha) = o(1/|\alpha|)$, as $|\alpha| \to \infty$, because of (3.4) and Theorem 1, since $f' \in L_1(-\infty,\infty)$ by assumption.

Remarks. Theorem 2(A) can be given a more general form. If m is a positive integer, $x^m f(x) \in L_1(-\infty < x < \infty)$, then $\hat{f}(\alpha)$ is continuously differentiable m times for $-\infty < \alpha < \infty$, and we have

$$(\hat{f})^{(m)}(\alpha) = \int_{-\infty}^{\infty} \{(ix)^m f(x)\} e^{i\alpha x}\,dx,$$

so that

$$|(\hat{f})^{(m)}(\alpha)| \leq \int_{-\infty}^{\infty} |(ix)^m f(x)|\,dx.$$

One has only to use Theorem 2(A) and induction on m.

Theorem 2(B) can also be similarly generalized. If f is continuously differentiable m times, and $f^{(r)}(x) \in L_1(-\infty < x < \infty)$, for $0 \leq r \leq m$, then

$$(f^{(m)})^{\wedge}(\alpha) = (-i\alpha)^m \hat{f}(\alpha),$$

so that

$$|\alpha^m \hat{f}(\alpha)| \leq \int_{-\infty}^{\infty} |f^{(m)}(x)|\,dx.$$

Thus one can roughly say that the faster f decreases, the more often is $\hat{f}$ differentiable (with bounded derivatives), and the more integrable derivatives f has, the faster $\hat{f}$ decreases.

The spaces $\mathcal{S}$ and $\mathcal{D}$ of L. Schwartz

A complex-valued function f(x) of the real variable x is said to belong to *Schwartz's space* $\mathcal{S}$, if f is differentiable infinitely often, and for any integers p,q,

$$x^p f^{(q)}(x) \to 0, \quad \text{as } |x| \to \infty,$$

$f^{(q)}$ denoting the qth derivative of f.

We note the following properties of $\mathcal{S}$.

(3.5) If $f \in \mathcal{S}$, then $x^\ell f^{(m)}(x)$ is bounded, and belongs to $L_1(-\infty,\infty)$, for any integers $\ell, m \geq 0$. For

$$|x^{\ell+2} f^{(m)}(x)| < M_0 < \infty$$

implies that

$$|x^\ell f^{(m)}(x)| < \frac{M}{1+x^2} \in L_1(-\infty,\infty).$$

In fact, f belongs to $L_p(-\infty,\infty)$ for every p such that $1 \leq p < \infty$.

(3.6) If $f \in \mathcal{S}$, then $(x^\ell f(x))^{(m)}$ is bounded, for any integers $\ell, m \geq 0$, and belongs to $L_1(-\infty,\infty)$.

This follows from (3.5), if we just use the rule for the differentiation of a product.

(3.7) If $f \in \mathcal{S}$, then $\hat{f} \in \mathcal{S}$.

First, if $f \in \mathcal{S}$, then $x^\ell f(x) \in L_1(-\infty < x < \infty)$, for *every* integer $\ell \geq 0$, so that (after Theorem 2(A)) $\hat{f}$ is differentiable infinitely often.

Secondly, if ℓ and m are positive integers, then by Theorem 2(A), and the remarks thereafter, we have

$$(\hat{f})^{(\ell)}(\alpha) = \int_{-\infty}^{\infty} [(ix)^{\ell} f(x)] e^{i\alpha x} dx,$$

so that by Theorem 2(B), and the remarks thereafter, we have

$$(3.71) \qquad |\alpha^m (\hat{f})^{(\ell)}(\alpha)| \leq \int_{-\infty}^{\infty} |\{(ix)^{\ell} f(x)\}^{(m)}| dx.$$

Thus

$$|(\hat{f})^{(\ell)}(\alpha)| = O(|\alpha|^{-m}), \quad \text{as } |\alpha| \to \infty,$$

for every $m \geq 0$, hence

$$|(\hat{f})^{(\ell)}(\alpha)| = o(|\alpha|^{-m}), \quad \text{as } |\alpha| \to \infty,$$

for every $m \geq 0$.

(3.8) We note that $e^{-x^2} \in S$, $\frac{1}{1+x^2} \notin S$, $e^{-|x|} \notin S$, the last not being differentiable at the origin.

(3.9) A concept of convergence, and therefore of continuity, can be introduced into the vector space S as follows. Given an infinite sequence (f_j) of functions, all belonging to S, we say that (f_j) *converges in* S to zero as $j \to \infty$, if for any integers $\ell, m \geq 0$, $\{x^{\ell} f_j^{(m)}(x)\}$ converges uniformly to zero on the line: $-\infty < x < \infty$.

(3.10) If $\{f_j(x)\}$ converges in S to zero as $j \to \infty$, then the sequence of Fourier transforms $(\hat{f}_j(\alpha))$ converges in S to zero.

For by (3.71) we have

$$|\alpha^m (\hat{f}_j)^{(\ell)}(x)| \leq \int_{-\infty}^{\infty} |\{(ix)^{\ell} f_j(x)\}^{(m)}| dx,$$

and the right-hand side tends to zero as $j \to \infty$, since

$$\int_{-\infty}^{\infty} |\{(ix)^{\ell} f_j(x)\}^{(m)}| dx = \int_{|x| \leq R} + \int_{|x| > R}, \quad R > 0$$

$$= I_1 + I_2, \text{ say},$$

where $I_2 \to 0$ as $R \to \infty$, uniformly in j, while $I_1 \to 0$ as $j \to \infty$ since the integrand there is bounded and converges uniformly to zero.

A continuous linear functional on S is known, after Schwartz, as a *tempered distribution*. A study of the theory of Fourier transforms of tempered distributions is outside the scope of the present text.

The space $\mathcal{D}$

The space S contains the vector subspace $\mathcal{D}$ of infinitely differentiable functions on $(-\infty,\infty)$ with *bounded supports*. The *support* of a function f defined on $(-\infty,\infty)$ is the closure of the set of points x at which $f(x) \neq 0$.

The function f defined by

$$f(x) = \begin{cases} 0, & \text{for } |x| \geq 1, \\ e^{-1/(1-x^2)}, & \text{for } |x| < 1, \end{cases}$$

belongs to $\mathcal{D}$. Its derivatives, of all orders, vanish at $x = \pm 1$.

We can introduce into $\mathcal{D}$ a concept of convergence.

(3.11) We say that an infinite sequence (f_j) of functions in $\mathcal{D}$ *converges in* $\mathcal{D}$ to a function f of $\mathcal{D}$ as $j \to \infty$, if the supports of (f_j) are all contained in the same bounded set (independently of j), and the sequence of derivatives $(f_j^{(m)})$, of any given order m, converges uniformly, as $j \to \infty$, to the derivative $f^{(m)}$ of f.

If (f_j) converges to zero in $\mathcal{D}$, then clearly (f_j) converges to zero in S as well.

A continuous linear functional on the space $\mathcal{D}$ is known, after Schwartz, as a *distribution*. One can study the Fourier transforms of particular classes of distributions, such as those with bounded supports, or those with point supports, though such a study is outside the scope of the present text.

We have already noted, and used, the fact that step-functions are *dense* in $L_p(-\infty,\infty)$ for $1 \leq p < \infty$ (cf. §1).

Given the function $\chi_{a,b}$ defined by

$$\chi_{a,b}(x) = \begin{cases} 1, & -\infty < a \leq x \leq b < +\infty, \\ 0, & \text{for } x < a,\ x > b, \end{cases}$$

we can approximate to it by a function $\omega_{a,b} \in \mathcal{D}$, such that for any given $\delta > 0$,

$$\| \chi_{a,b} - \omega_{a,b} \|_p < \delta.$$

Let

$$\omega(x) = \begin{cases} 1, & \text{for } a < x < b, \\ 0, & \text{for } x < a-\varepsilon, \\ 0, & \text{for } x > b+\varepsilon, \end{cases} \qquad \varepsilon > 0, \tag{3.12}$$

while

$$\omega(x) = \frac{1}{c_1} \int_{a-\varepsilon}^{x} e^{\frac{1}{(y-a)(y-a+\varepsilon)}} dy, \qquad \text{for } a-\varepsilon \le x \le a,$$

$$= \frac{1}{c_2} \int_{x}^{b+\varepsilon} e^{\frac{1}{(y-b)(y-b-\varepsilon)}} dy, \qquad \text{for } b \le x \le b+\varepsilon,$$

where

$$c_1 = \int_{a-\varepsilon}^{a} e^{\frac{1}{(y-a)(y-a+\varepsilon)}} dy, \qquad c_2 = \int_{b}^{b+\varepsilon} e^{\frac{1}{(y-b)(y-b-\varepsilon)}} dy.$$

Then $\omega(x) = 1$, for $x = a$ and $x = b$, while $\omega(x) = 0$ for $x = a-\varepsilon$, and $x = b+\varepsilon$; and the derivatives of ω, of all orders, vanish at $x = a-\varepsilon$, a, b, $b+\varepsilon$. If ε is chosen sufficiently small, the graph of the function ω is as shown in the figure:

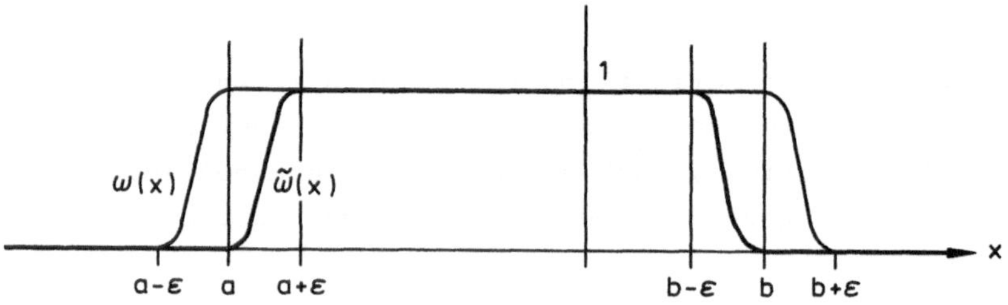

Clearly $\omega \in L_p(-\infty,\infty)$, for every p such that $1 \le p < \infty$, and

$$\| \chi_{a,b} - \omega \|_p < (2\varepsilon)^{1/p}.$$

It follows that $\mathcal{D}$ is a dense subspace of $L_p(-\infty,\infty)$, $1 \leq p < \infty$, and since $S \supset \mathcal{D}$, S is also a dense subspace of $L_p(-\infty,\infty)$.

Since $\omega \in \mathcal{D} \subset S$, we have $\hat{\omega} \in S$.

We can similarly construct an infinitely differentiable function $\tilde{\omega} = \tilde{\omega}_{a,b}$, such that

$$\tilde{\omega}(x) = \begin{cases} 1, & \text{for } a+\varepsilon \leq x \leq b-\varepsilon, \\ 0, & \text{for } x \leq a,\ x \geq b, \end{cases}$$

so that

$$\tilde{\omega}(x) < \chi_{a,b}(x) < \omega(x),$$

and

$$\int_{-\infty}^{\infty} \{\omega(x) - \tilde{\omega}(x)\}dx < 4\varepsilon.$$

These auxiliary functions will come in useful in the proof of the Central Limit Theorem in §10.

§4. Localization, Mellin transforms

If $f(x) \in L_1(-\infty < x < \infty)$, it does *not* follow that the Fourier transform $\hat{f}(\alpha)$ belongs to $L_1(-\infty < \alpha < \infty)$, as in Example 1 of §1. We can, however, find a simple condition, such that *at a given point* x, we can "invert" the Fourier transform; that is to say, obtain the relation

$$f(x) = \frac{1}{2\pi} \int_{-\infty}^{\infty} f(\alpha) e^{-i\alpha x} d\alpha,$$

the integral being defined as (a Cauchy principal value)

$$\lim_{R\to\infty} \int_{-R}^{R} \hat{f}(\alpha) e^{-i\alpha x} d\alpha, \quad R > 0.$$

Let $f \in L_1(-\infty,\infty)$, and

(4.1) $$S_R(x) = \frac{1}{2\pi}\int_{-R}^{R} e^{-ix\alpha}\hat{f}(\alpha)\,d\alpha, \quad 0 < R < \infty.$$

Then we have

$$S_R(x) = \frac{1}{2\pi}\int_{-R}^{R} e^{-ix\alpha}\left(\int_{-\infty}^{\infty} f(t)e^{i\alpha t}dt\right)d\alpha,$$

and since the repeated integral is absolutely convergent, the order of integration can be interchanged, so that

$$S_R(x) = \frac{1}{2\pi}\int_{-\infty}^{\infty} f(t)\left(\int_{-R}^{R} e^{-i\alpha(x-t)}\,d\alpha\right)dt$$

$$= \frac{1}{2\pi}\int_{-\infty}^{\infty} f(t)\left[\frac{i}{x-t}\left\{e^{-iR(x-t)} - e^{iR(x-t)}\right\}\right]dt$$

$$= \frac{1}{\pi}\int_{-\infty}^{\infty} f(x-t)\,\frac{\sin Rt}{t}\,dt.$$

Hence we have

(4.2) $$S_R(x) = \frac{1}{\pi}\int_0^{\infty}\left\{f(x+t) + f(x-t)\right\}\frac{\sin Rt}{t}\,dt,$$

or

$$S_R(x) - f(x) = \frac{2}{\pi}\int_0^{\infty}\left\{\frac{f(x+t) + f(x-t)}{2} - f(x)\right\}\frac{\sin Rt}{t}\,dt,$$

since, for $R > 0$,

(4.3) $$\int_0^{\infty}\frac{\sin Rt}{t}\,dt = \frac{\pi}{2},$$

the integral being convergent though *not* absolutely. Thus we obtain the following

(4.4) <u>*Lemma*</u>. *If* $f \in L_1(-\infty,\infty)$, *and*

(4.1) $$S_R(x) = \frac{1}{2\pi}\int_{-R}^{R}\hat{f}(\alpha)e^{-i\alpha x}\,d\alpha, \quad 0 < R < \infty,$$

and

(4.5) $$g_x(t) = \frac{1}{2}\,[f(x+t) + f(x-t) - 2f(x)],$$

then

$$S_R(x) - f(x) = \frac{2}{\pi} \int_0^\infty g_x(t) \frac{\sin Rt}{t} dt.$$

This lemma will enable us to prove that the convergence of $S_R(x)$, as $R \to \infty$, to $f(x)$ at a given point x, depends only on the behaviour of the function in a neighbourhood of that point. This is usually referred to as *Riemann's localization theorem*, and is contained in the following two theorems.

Theorem 4. *If* $f \in L_1(-\infty,\infty)$, $x \in (-\infty,\infty)$, *and* $g_x(t)$ *is defined as in (4.5), and there exists a* $\delta > 0$, *such that*

$$\int_0^\delta \frac{|g_x(t)|}{t} dt < \infty, \tag{4.6}$$

then we have

$$\lim_{R\to\infty} S_R(x) = f(x),$$

where $S_R(x)$ *is defined as in (4.1).*

Proof. For any fixed $\delta > 0$, we have

$$S_R(x) - f(x) = \frac{2}{\pi} \left[\int_0^\delta + \int_\delta^\infty\right] \frac{g_x(t)}{t} \sin Rt \, dt = I_1 + I_2, \text{ say.}$$

Since $g_x(t)/t$ is absolutely integrable in $(0,\delta)$ by hypothesis, we have

$$I_1 = \varepsilon(\delta) \to 0, \quad \text{as } \delta \downarrow 0;$$

while Theorem 1 implies that for fixed $\delta > 0$,

$$\int_\delta^\infty \frac{f(x+t)}{t} \sin Rt \, dt \to 0, \quad \text{and} \int_\delta^\infty \frac{f(x-t)}{t} \sin Rt \, dt \to 0,$$

as $R \to \infty$. The integral

$$\int_\delta^\infty f(x) \frac{\sin Rt}{t} dt = f(x) \int_{\delta R}^\infty \frac{\sin T}{T} dT \to 0, \quad \text{as } R \to \infty$$

(cf. (4.3)). Hence $I_2 \to 0$, as $R \to \infty$, which proves the theorem.

Remark. Condition (4.6) is satisfied at any point x at which the function f has a finite derivative, or satisfies just a Lipschitz condition of order α, namely: $|f(x+h) - f(x)| = O(|h|^\alpha)$, $0 < \alpha < 1$.

The next theorem gives a sufficient condition for the convergence of the integral $S_R(x)$ as $R\to\infty$, not only *at* the point x, but in an interval containing the point.

Theorem 5. If $f\in L_1(-\infty,\infty)$*, and* f *is of bounded variation in a neighbourhood of the point* x*, then we have*

$$\lim_{R\to\infty} S_R(x) = \tfrac{1}{2}[f(x+0) + f(x-0)],$$

where $S_R(x)$ *is defined as in (4.1).*

Proof. If for a given δ, $\delta > 0$, the function g is of bounded variation in the interval $[0,\delta]$, then we shall prove that

$$\lim_{R\to\infty} \frac{2}{\pi}\int_0^\delta g(t)\,\frac{\sin Rt}{t}\,dt = g(0+). \tag{4.7}$$

Since g can be expressed as the difference of two bounded, monotone increasing functions, it suffices to prove (4.7) on the assumption that g is a bounded, monotone increasing function. We may further assume that $g(0+) = 0$, for if $g(0+) \neq 0$, we set $G(t) = g(t) - g(0+)$, so that $G(0+) = 0$, and *if* (4.7) holds with G in place of g, then we have

$$\lim_{R\to\infty}\int_0^\delta G(t)\,\frac{\sin Rt}{t}\,dt = G(0+) = 0,$$

which implies, in turn, that

$$\frac{2}{\pi}\int_0^\delta g(t)\,\frac{\sin Rt}{t}\,dt = \frac{2}{\pi}\int_0^\delta G(t)\,\frac{\sin Rt}{t}\,dt + \frac{2}{\pi}\int_0^\delta g(0+)\,\frac{\sin Rt}{t}\,dt$$

$$\to 0 + g(0+), \quad \text{as } R\to\infty,$$

because of (4.3), which proves (4.7).

We assume therefore that $g(0+) = 0$, and that g is a bounded, monotone increasing function. Given $\varepsilon > 0$, therefore, we can find η, such that $0 < \eta < \delta$, and $|g(t)| \le \varepsilon$ for $0 < t \le \eta$. We now apply the second mean-value theorem, which states that if f is integrable on the finite interval (a,b) and φ bounded and monotone in (a,b), then we have

$$\int_a^b f(x)\varphi(x)\,dx = \varphi(a+0)\int_a^\xi f(x)\,dx + \varphi(b-0)\int_\xi^b f(x)\,dx,$$

for some $\xi \in [a,b]$. On setting $\varphi = g$, $f(t) = \frac{\sin Rt}{t}$, we see that there exists $\xi \in [0,\eta]$, such that

$$\int_0^\eta g(t) \frac{\sin Rt}{t} dt = g(\eta-0) \int_\xi^\eta \frac{\sin Rt}{t} dt = g(\eta-0) \int_{\xi R}^{\eta R} \frac{\sin t}{t} dt.$$

Hence

$$\left|\int_0^\eta g(t) \frac{\sin Rt}{t} dt\right| \leq |g(\eta-0)| \cdot M \leq \varepsilon M < \infty,$$

where ε is an arbitrary positive number, and M is independent of ξ,η, and R (cf. (1.18)). And we have

$$\left|\int_0^\delta g(t) \frac{\sin Rt}{t} dt\right| \leq \varepsilon M + \left|\int_\eta^\delta g(t) \frac{\sin Rt}{t} dt\right|.$$

Since g is bounded and measurable on $[0,\delta]$, the function $g(t)/t$ is integrable on $[\eta,\delta]$, so that Theorem 1 gives

$$\lim_{R\to\infty} \int_\eta^\delta \frac{g(t)}{t} \sin Rt \, dt = 0,$$

and hence

$$\limsup_{R\to\infty} \left|\int_0^\delta g(t) \frac{\sin Rt}{t} dt\right| \leq \varepsilon M,$$

or

$$\lim_{R\to\infty} \int_0^\delta g(t) \frac{\sin Rt}{t} dt = 0 = g(0+).$$

We shall now use (4.7) to prove the theorem.

Given x, we choose δ, such that $\delta > 0$, and δ is so small that f is of bounded variation in $[x-\delta, x+\delta]$. By (4.2) we have

$$S_R(x) = \frac{1}{\pi} \left[\int_0^\delta + \int_\delta^\infty\right] \left\{\frac{f(x+t) + f(x-t)}{2}\right\} \frac{\sin Rt}{t} dt$$

$$= I_1 + I_2, \text{ say.}$$

Since $\{f(x+t) + f(x-t)\}/t$ is integrable, as a function of t, on the interval $\delta \leq t < \infty$, Theorem 1 implies that $I_2 \to 0$, as $R \to \infty$. By (4.7), however, we have

$$I_1 \to \frac{1}{2}\,[f(x+0) + f(x-0)], \text{ as } R\to\infty,$$

and Theorem 5 follows.

Remarks. The criteria for the convergence of $S_R(x)$ to $f(x)$, as $R\to\infty$, given by Theorems 4 and 5 are not comparable. The function f defined by $f(x) = x\sin(1/x)$, for $0 < x \le \frac{1}{\pi}$; $f(x) = 0$, for $x \le 0$; $f(x) = 0$, for $x > \frac{1}{\pi}$; satisfies condition (4.6) of Theorem 4 at $x = 0$, but it is not of bounded variation in any neighbourhood of the origin.

The function f defined by $f(x) = 1/\log(1/x)$ for $0 < x < \frac{1}{e}$, $f(x) = 1/(ex)^2$ for $x \ge \frac{1}{e}$, $f(x) = f(-x)$, is of bounded variation in a neighbourhood of the origin (since it is bounded and monotone), but does not satisfy condition (4.6) of Theorem 4 at $x = 0$. Both the functions belong, of course, to $L_1(-\infty,\infty)$.

Mellin transforms

Theorem 5 leads us, by a change of variable, to *Mellin transforms* and *Mellin inversion.*

Theorem 5'. Let $y^{\sigma-1}f(y) \in L_1(0 < y < \infty)$, *for* σ *real, and let* $f(y)$ *be of bounded variation in a neighbourhood of the point* $y = x$. *If*

$$F(s) = \int_0^\infty f(y)y^{s-1}dy, \qquad s = \sigma + it, \quad i = \sqrt{-1}, \tag{4.8}$$

then

$$\frac{1}{2\pi i}\lim_{T\to\infty}\int_{\sigma-iT}^{\sigma+iT} F(s)x^{-s}dx = \frac{1}{2}\,[f(x+0) + f(x-0)]. \tag{4.9}$$

Proof. The hypothesis on f ensures the existence of the integral in (4.8). As hitherto f is a complex-valued function of the real variable x; on the other hand, s is complex with real part σ and imaginary part t. If we make the substitution $x = e^y$ in (4.8), we get

$$F(\sigma+it) = \int_{-\infty}^{\infty} f(e^y)e^{\sigma y}\, e^{ity}dy,$$

so that $F(\sigma+it)$, considered as a function of t, $-\infty < t < \infty$, may be looked upon as the Fourier transform of $f(e^y)e^{\sigma y} \in L_1(-\infty < y < \infty)$. By Theorem 5 we get

$$\lim_{R\to\infty} \frac{1}{2\pi} \int_{-R}^{R} F(\sigma+it)e^{-ity}dt = \frac{1}{2}\,[g(y+0) + g(y-0)],$$

where $g(y) = f(e^y)\,e^{\sigma y}$, provided that g is of bounded variation in a neighbourhood of the point y, or

$$\lim_{R\to\infty} \frac{1}{2\pi} \int_{-R}^{R} F(\sigma+it)\; e^{-(\sigma+it)y}dt = e^{-\sigma y}\,\frac{1}{2}\,[g(y+0) + g(y-0)].$$

On setting $x = e^y$, we get

$$\lim_{R\to\infty} \frac{1}{2\pi i} \int_{\sigma-iR}^{\sigma+iR} F(s)x^{-s}ds = \frac{1}{2}\,[f(x+0) + f(x-0)],$$

which is (4.9).

Similarly we have also the following

Theorem 5″. *Let* $F(\sigma+iu) \in L_1(-\infty < u < \infty)$, *and let* F *be of bounded variation, as a function of* u, *in a neighbourhood of the point* $u = t$. *If*

$$f(x) = \frac{1}{2\pi i} \int_{\sigma-i\infty}^{\sigma+i\infty} F(s)x^{-s}ds, \qquad s = \sigma+it,$$

then

$$\lim_{R\to\infty} \int_{1/R}^{R} f(x)x^{\sigma+it-1}dx = \frac{1}{2}\,[F\{\sigma+i(t+0)\} + F\{\sigma+i(t-0)\}].$$

The function F in (4.8) is usually referred to as the *Mellin transform* of f; and (4.9) is referred to as the *Mellin inversion formula*. We note that the Mellin transform is just another version of the Fourier transform obtained by a change of variable.

Examples

1. The integral representation for the gamma-function given by

$$\Gamma(s) = \int_0^\infty e^{-x}x^{s-1}dx, \quad \sigma = \mathrm{Re}\ s > 0,$$

shows that in Theorem 5' if $f(x) = e^{-x}$, then $F(s) = \Gamma(s)$, for $\sigma > 0$. Thus $\Gamma(s)$, for $\sigma > 0$, is the Mellin transform of e^{-x}, $0 < x < \infty$. And we have

$$e^{-x} = \frac{1}{2\pi i} \int_{\sigma-i\infty}^{\sigma+i\infty} \Gamma(s)x^{-s}ds, \quad \sigma > 0,\ x > 0.$$

2. The series $\sum_{n=1}^{\infty} n^{-s}$, $s = \sigma+it$, converges for $\sigma > 1$, and the sum-function $\zeta(s)$ is known as the *Riemann zeta-function*. If in Theorem 5' $f(x) = 1/(e^x-1)$, then $F(s) = \Gamma(s)\zeta(s)$, for $\sigma > 1$.

To see this we note that for any integer n, $n \geq 1$, we have

$$\Gamma(s)n^{-s} = \int_0^{\infty} e^{-nx}\, x^{s-1}dx, \quad \text{for } \sigma > 0,$$

and since

$$\sum_{n=1}^{\infty} \int_0^{\infty} |x^{s-1}e^{-nx}|dx = \sum_{n=1}^{\infty} \int_0^{\infty} e^{-nx}\, x^{\sigma-1}dx = \sum_{n=1}^{\infty} \Gamma(\sigma)n^{-\sigma} < \infty,$$

for $\sigma > 1$, we have

$$\Gamma(s)\zeta(s) = \sum_{n=1}^{\infty} \int_0^{\infty} x^{s-1}\, e^{-nx}dx = \int_0^{\infty} x^{s-1} \sum_{n=1}^{\infty} e^{-nx}dx = \int_0^{\infty} \frac{x^{s-1}}{e^x-1}\, dx,$$

for $\sigma > 1$. Thus $\Gamma(s)\zeta(s)$, for $\sigma > 1$, is the Mellin transform of $1/(e^x-1)$, $0 < x < \infty$, and we deduce that

$$\frac{1}{e^x-1} = \frac{1}{2\pi i} \int_{\sigma-i\infty}^{\sigma+i\infty} \Gamma(s)\zeta(s)x^{-s}ds, \quad \sigma > 1, \ x > 0.$$

3. Let L(s) denote one of Dirichlet's L-functions, defined by the series

$$L(s) = \frac{1}{1^s} - \frac{1}{3^s} + \frac{1}{5^s} - \ldots., \quad \text{for } \sigma > 0;$$

then $\Gamma(s)L(s)$, for $\sigma > 0$, is the Mellin transform of $\dfrac{1}{e^x+e^{-x}}$, $0 < x < \infty$, and we deduce that

$$\frac{1}{e^x+e^{-x}} = \frac{1}{2\pi i} \int_{\sigma-i\infty}^{\sigma+i\infty} \Gamma(s)L(s)x^{-s}ds, \quad \sigma > 0, \ x > 0.$$

§5. Fourier series and Poisson's summation formula

If $f(x) \in L_1(0 \leq x \leq 2\pi)$, and $f(x+2\pi) = f(x)$, for $-\infty < x < \infty$, the *Fourier series* of f is defined to be

(5.1) $$\sum_{\nu=-\infty}^{\infty} c_\nu e^{i\nu x}, \quad 0 \leq x \leq 2\pi,$$

where the *Fourier coefficient* c_ν is given by

(5.2) $$c_\nu = \frac{1}{2\pi}\int_0^{2\pi} f(x)e^{-i\nu x}dx .$$

If $g_x(t) = \frac{1}{2}[f(x+t) + f(x-t) - 2f(x)]$, and there exists a $\delta > 0$, such that $\int_0^\delta |g_x(t)|t^{-1}dt < \infty$, then the Fourier series of f *at the point* x converges to sum f(x). This is the well-known criterion of convergence due to Dini, of which Theorem 4 is the analogue for Fourier transforms.

If on the other hand, f is of bounded variation in $(0,2\pi)$, then at every point x_0 the Fourier series converges to $\frac{1}{2}[f(x_0+0) + f(x_0-0)]$. In particular, the series converges to f(x) at every point of continuity of f. If further f is continuous at every point of a closed interval, then the series converges *uniformly* in that interval. This is the well-known criterion of convergence due to Dirichlet and Jordan, of which Theorem 5 is the analogue for Fourier transforms.

The following lemma establishes a simple connexion between Fourier transforms of functions in $L_1(-\infty,\infty)$ and the Fourier series of related periodic functions, of period 2π, belonging to $L_1(0,2\pi)$.

(5.3) Lemma. *If* $f(x) \in L_1(-\infty < x < \infty)$, *then the series* $\sum_{k=-\infty}^{\infty} f(x + 2k\pi)$ *converges absolutely for almost all* x *in* $(0,2\pi)$ *and its sum* F(x) *belongs to* $L_1(0,2\pi)$, *with* $F(x+2\pi) = F(x)$ *for all real* x. *If* C_ν *denotes the Fourier coefficient of* F, *then*

(5.4) $$C_\nu \equiv \frac{1}{2\pi}\int_0^{2\pi} F(x)e^{-i\nu x}dx = \frac{1}{2\pi}\int_{-\infty}^{\infty} f(x)e^{-i\nu x}dx \equiv \frac{1}{2\pi}\hat{f}(-\nu).$$

Proof. We have

$$\sum_{k=-\infty}^{\infty}\int_0^{2\pi}|f(x+2k\pi)|dx \equiv \lim_{N\to\infty}\sum_{k=-N}^{N}\int_0^{2\pi}|f(x+2k\pi)|dx$$

$$= \lim_{N\to\infty}\left(\sum_{k=-N}^{N}\int_{2k\pi}^{(2k+2)\pi}|f(y)|dy\right)$$

$$= \lim_{N\to\infty}\int_{-2N\pi}^{(2N+2)\pi}|f(y)|dy = \int_{-\infty}^{\infty}|f(y)|dy < \infty .$$

It follows by Lebesgue's theorem on monotone convergence that

$$\int_0^{2\pi} \sum_{k=-\infty}^{\infty} |f(x+2k\pi)|\,dx = \sum_{k=-\infty}^{\infty} \int_0^{2\pi} |f(x+2k\pi)|\,dx < \infty,$$

hence $\sum_{k=-\infty}^{\infty} f(x+2k\pi)$ converges absolutely for almost all x in $(0,2\pi)$, and if $F_N(x) = \sum_{k=-N}^{N} f(x+2k\pi)$, then $\lim_{N\to\infty} F_N(x) = F(x)$, where $F \in L_1(0,2\pi)$, and $F(x+2\pi) = F(x)$. The νth Fourier coefficient of F is given by

$$\frac{1}{2\pi}\int_0^{2\pi} F(x)e^{-i\nu x}dx \equiv \frac{1}{2\pi}\int_0^{2\pi} \lim_{N\to\infty} F_N(x)e^{-i\nu x}dx$$

$$= \lim_{N\to\infty} \frac{1}{2\pi}\int_{-2N\pi}^{(2N+2)\pi} f(x)e^{-i\nu x}dx = \frac{1}{2\pi}\int_{-\infty}^{\infty} f(x)e^{-i\nu x}dx,$$

since

$$|F_N(x)| < \sum_{k=-\infty}^{\infty} |f(x+2k\pi)| \in L_1(0,2\pi).$$

<u>*Theorem 6*</u>. *Let* $f \in L_1(-\infty,\infty)$, *and be of bounded variation on* $(-\infty,\infty)$, *and let* $f(x) = \frac{1}{2}[f(x+0) + f(x-0)]$ *for all* x *in* $(-\infty,\infty)$. *Then we have*

$$\sum_{k=-\infty}^{\infty} f(2k\pi) = \sum_{\nu=-\infty}^{\infty} \frac{1}{2\pi}\int_{-\infty}^{\infty} f(x)e^{-i\nu x}dx \equiv \lim_{N\to\infty} \sum_{\nu=-N}^{N} (\quad). \tag{5.5}$$

<u>Proof</u>. Let v_k denote the total variation of f in the interval $I_k = (2k\pi,(2k+2)\pi)$, $k = 0, \pm 1, \pm 2, \ldots$. The series $\sum_{k=-\infty}^{\infty} f(x+2k\pi)$ converges absolutely at some point x_0 in I_0, and

$$\sum_{|k|\geq N} |f(x+2k\pi)| \leq \sum_{|k|\geq N} |f(x_0+2k\pi)| + \sum_{|k|\geq N} |f(x+2k\pi) - f(x_0+2k\pi)|,$$

where

$$|f(x+2k\pi) - f(x_0+2k\pi)| \leq v_k, \quad \text{for } x \in I_0.$$

Since $\sum_{k=-\infty}^{\infty} v_k = \lim_{n\to\infty} \sum_{k=-n}^{n} v_k < \infty$, by hypothesis, the series $\sum_{k=-\infty}^{\infty} f(x+2k\pi)$ converges absolutely, and uniformly, in I_0 to sum F(x), say, which is of bounded variation, and such that $F(x) = \frac{1}{2}[F(x+0) + F(x-0)]$, $F(x+2\pi) = F(x)$. By the Dirichlet-Jordan test mentioned above, the

Fourier series of F, say $\sum_{\nu=-\infty}^{\infty} c_\nu \, e^{+i\nu x}$, converges to F(x), so that

$$\sum_{k=-\infty}^{\infty} f(x+2k\pi) = F(x) = \sum_{\nu=-\infty}^{\infty} c_\nu \, e^{i\nu x},$$

and at the point x = 0, we have, by Lemma (5.3),

$$\sum_{k=-\infty}^{\infty} f(2k\pi) = \sum_{\nu=-\infty}^{\infty} c_\nu = \sum_{\nu=-\infty}^{\infty} \frac{1}{2\pi} \int_{-\infty}^{\infty} f(x) e^{-i\nu x} dx.$$

Remarks. Formula (5.5) is referred to as Poisson's summation formula. The conditions for its validity can, of course, be relaxed. A more symmetric form can be obtained by modifying the definition of Fourier transform. If we write $f(x) = g\left(\frac{ax}{2\pi}\right)$, where $a > 0$, and $ab = 2\pi$, and *define*

$$\check{F}[f](\alpha) = \frac{1}{\sqrt{2\pi}} \int_{-\infty}^{\infty} f(x) e^{-i\alpha x} dx, \qquad \left(= \frac{1}{\sqrt{2\pi}} \hat{f}(-\alpha)\right) \tag{5.6}$$

then (5.5) takes the form

$$\sqrt{a} \sum_{k=-\infty}^{\infty} g(ak) = \sqrt{b} \sum_{\nu=-\infty}^{\infty} \check{F}[g](b\nu), \qquad ab = 2\pi, \; a > 0, \tag{5.7}$$

where $g \in L_1(-\infty,\infty)$.

Examples

1. If we take $g(x) = e^{-x^2}$, $a = (\pi t)^{1/2}$, $t > 0$, $b = (2\sqrt{\pi})/\sqrt{t}$, as we may, and use Example 6 of §1, we obtain from (5.7) the theta-relation

$$\sum_{k=-\infty}^{\infty} e^{-\pi k^2 t} = \frac{1}{\sqrt{t}} \sum_{k=-\infty}^{\infty} e^{-\pi k^2/t}, \qquad t > 0. \tag{5.8}$$

2. If we take $g(x) = e^{-|x|}$, and use Example 4 of §1, we obtain from (5.7) the formula

$$\sqrt{a} \sum_{k=-\infty}^{\infty} e^{-|k|a} = \sqrt{\left(\frac{2b}{\pi}\right)} \sum_{n=-\infty}^{\infty} \frac{1}{1+n^2b^2}, \quad ab = 2\pi, \; a > 0. \tag{5.9}$$

§6. The uniqueness theorem

If the Fourier transform $\hat{f}$ of a function $f \in L_1(-\infty,\infty)$ vanishes everywhere, then the function itself must vanish almost everywhere. This can be proved in many different ways, as we shall see later. We can prove it, at this stage, by using the infinitely differentiable function ω which vanishes outside a finite interval, introduced in (3.12), and applying Theorem 4.

Theorem 7. If $f(x) \in L_1(-\infty < x < \infty)$, *and* $\hat{f}$ *denotes the Fourier transform of* f, *and* $\hat{f}(\alpha) = 0$ *for every* α *such that* $-\infty < \alpha < \infty$, *then* $f(x) = 0$ *for almost all* x, $-\infty < x < \infty$.

Proof. Given real numbers $c > 0$ and $\varepsilon > 0$, let

$$\omega_{c,\varepsilon}(x) = \begin{cases} 0, & \text{for } x < -c-\varepsilon, \text{ and } x > c+\varepsilon, \\ 1, & \text{for } -c < x < c, \end{cases}$$

and let $\omega_{c,\varepsilon}(x)$ be infinitely differentiable for $-\infty < x < \infty$. Its derivatives, of all orders greater than zero, vanish at $x = -c-\varepsilon$, $-c$, $+c$, $c+\varepsilon$. Such a function exists; we have only to take $a = -c$, $b = +c$ in (3.12). Obviously we have

$$\hat{\omega}_{c,\varepsilon}(\alpha) = \int_{-\infty}^{\infty} \omega_{c,\varepsilon}(x)e^{i\alpha x}dx,$$

and on integrating this by parts sufficiently often, we see that $\hat{\omega}_{c,\varepsilon}(\alpha) = O(|\alpha|^{-k})$, as $|\alpha| \to \infty$, for *any* integer $k \geq 1$, and hence, in particular, $\hat{\omega}_{c,\varepsilon}(\alpha) \in L_1(-\infty < \alpha < \infty)$. Because $\omega_{c,\varepsilon}(x)$ has a finite derivative at every point x, $-\infty < x < \infty$, assumption (4.6) of Theorem 4 is satisfied everywhere, and we can conclude that

$$\omega_{c,\varepsilon}(x) = \frac{1}{2\pi} \int_{-\infty}^{\infty} \hat{\omega}_{c,\varepsilon}(\alpha)e^{-ix\alpha}d\alpha, \qquad -\infty < x < \infty,$$

where the integral converges *absolutely*, since $\hat{\omega}_{c,\varepsilon} \in L_1(-\infty,\infty)$. By the composition rule (1.13), and (1.9), we obtain

$$\int_{-\infty}^{\infty} f(y)\, \omega_{c,\varepsilon}(x-y)\,dy = \frac{1}{2\pi} \int_{-\infty}^{\infty} \hat{f}(\alpha)\, (\hat{\omega}_{c,\varepsilon}(\alpha)e^{-i\alpha x})\,d\alpha.$$

Since $\hat{f}(\alpha) = 0$, for every α, we obtain

(6.1) $$\int_{-\infty}^{\infty} f(y)\omega_{c,\varepsilon}(x-y)\,dy = 0,$$

which holds for every $c > 0$. From the definition of $\omega_{c,\varepsilon}$, we have

$$\int_{-\infty}^{\infty} f(y)\omega_{c,\varepsilon}(x-y)\,dy = \left(\int_{x-c-\varepsilon}^{x-c} + \int_{x-c}^{x+c} + \int_{x+c}^{x+c+\varepsilon}\right) f(y)\omega_{c,\varepsilon}(x-y)\,dy,$$

where

$$\left|\int_{x-c-\varepsilon}^{x-c} f(y)\omega_{c,\varepsilon}(x-y)\,dy\right| \le \int_{x-c-\varepsilon}^{x-c} |f(y)|\cdot 1\,dy \to 0, \quad \text{as } \varepsilon \downarrow 0,$$

and similarly also

$$\left|\int_{x+c}^{x+c+\varepsilon} f(y)\omega_{c,\varepsilon}(x-y)\,dy\right| \to 0, \quad \text{as } \varepsilon \downarrow 0,$$

while

$$\int_{x-c}^{x+c} f(y)\omega_{c,\varepsilon}(x-y)\,dy = \int_{x-c}^{x+c} f(y)\,dy.$$

By (6.1) it follows that for arbitrary x,

$$\int_{x-c}^{x+c} f(y)\,dy = 0,$$

for *every* $c > 0$; that is to say, $\int_{\alpha}^{\beta} f(y)dy = 0$, for arbitrary α and β, which implies that $f(x) = 0$ for almost all x, $-\infty < x < \infty$.

Remark. The above proof makes use of the infinitely differentiable function ω together with the validity of "Fourier inversion", namely

$$\omega(x) = \frac{1}{2\pi}\int_{-\infty}^{\infty} \hat{\omega}(\alpha)e^{-i\alpha x}\,d\alpha.$$

We shall presently see that if *both* f and $\hat{f}$ belong to $L_1(-\infty,\infty)$ such an inversion holds almost everywhere, from which Theorem 6 would follow at once.

(6.2) *Corollary. If* $f_1, f_2 \in L_1(-\infty,\infty)$, *and* $\hat{f}_1 = \hat{f}_2$ *everywhere, then* $f_1 = f_2$ *almost everywhere.*

§7. Pointwise summability

Examples given in §1 show that if $f \in L_1(-\infty,\infty)$, it does *not* necessarily follow that the Fourier transform $\hat{f}$ of f also belongs to $L_1(-\infty,\infty)$, so that the integral - referred to sometimes as a *Fourier integral* -

$$\frac{1}{2\pi}\int_{-\infty}^{\infty} \hat{f}(\alpha)e^{-i\alpha x}d\alpha$$

may not exist as a Lebesgue integral, or even as a Cauchy principal value. We can, however, introduce into the integrand a function $K(\alpha)$, called a *kernel*, or a *convergence factor*, or a *summability factor*, and formulate general conditions on K, and on *its* Fourier transform, to secure the relation

$$\lim_{R\to\infty}\int_{-R}^{R} \hat{f}(\alpha)K(\tfrac{\alpha}{R})\, e^{-i\alpha x}d\alpha = f(x),$$

for almost every x.

Theorem 8. If $K \in L_1(-\infty,\infty)$, K is even, and $\hat{K} \equiv H$, and $R > 0$, then we have, for every $f \in L_1(-\infty,\infty)$, the formula

$$(7.1)\qquad \int_{-\infty}^{\infty} \hat{f}(\alpha)\, K_R(\alpha)e^{-i\alpha x}d\alpha = \int_{-\infty}^{\infty} f(t)\, H_R(x-t)dt \equiv (f*H_R)(x),$$

where $K_R(\alpha) = K(\alpha/R)$, *and* $H_R(x) = R\,H(Rx)$.

If we assume further that $\frac{1}{2\pi}\int_{-\infty}^{\infty} H(t)dt = 1$, *then we have the formula*

$$(7.2)\qquad \frac{1}{2\pi}(f*H_R)(x) - f(x) = \frac{1}{\pi}\int_{0}^{\infty} g_x(t)\; RH(Rt)dt,$$

where

$$(7.3)\qquad g_x(t) = \frac{1}{2}\,[f(x+t) + f(x-t) - 2f(x)],\; 0 \le t < \infty.$$

Proof. If $f \in L_1(-\infty,\infty)$, the composition rule (1.13) gives

$$(7.4)\qquad \int_{-\infty}^{\infty} \hat{f}(\alpha)K_R(\alpha)e^{-ix\alpha}d\alpha = \int_{-\infty}^{\infty} f(x+t)H_R(t)dt = \int_{-\infty}^{\infty} f(y)H_R(y-x)dy.$$

If we assume, in addition, that K is even, then H is even (cf. (1.10)) and the last integral equals the convolution $(f*H_R)(x)$, giving (7.1).

If we assume further that $\frac{1}{2\pi}\int_{-\infty}^{\infty} H(t)dt = 1$, then

$$\frac{1}{2\pi}(f*H_R)(x) - f(x) = \frac{1}{2\pi}\int_{-\infty}^{\infty} [f(x+t) - f(x)]\ RH(Rt)dt$$

$$= \frac{1}{\pi}\int_{0}^{\infty} g_x(t)\ RH(Rt)dt,$$

giving (7.2).

Formulas (7.1) and (7.2) can be used to formulate conditions under which

$$\frac{1}{2\pi}\ (f*H_R)(x) \to f(x),$$

as $R \to \infty$.

<u>Theorem 9</u>. *Let* $f \in L_1(-\infty,\infty)$, *and for each* $x \in (-\infty,\infty)$ *let*

$$g_x(t) = \frac{1}{2}\ [f(x+t) + f(x-t) - 2f(x)], \quad 0 \le t < \infty.$$

Let $K \in L_1(-\infty,\infty)$, K *even*, $\hat{K} \equiv H \in L_1(-\infty,\infty)$, *and* $\frac{1}{2\pi}\int_{-\infty}^{\infty} H(t)dt = 1$. *Let* $H(t)$ *be monotone decreasing for* $0 \le t < \infty$. *Then*

$$\frac{1}{2\pi}\ (f*H_R)(x) \to f(x), \quad \text{as } R \to \infty, \qquad \left(H_R(x) = RH(Rx),\ R > 0\right) \tag{7.5}$$

at every point x *at which*

$$\lim_{h \to 0} \frac{1}{h}\int_{0}^{h} g_x(t)dt = 0. \tag{7.6}$$

In particular, (7.5) holds at every point x *at which* f *is continuous; and uniformly over every closed interval of points of continuity of* f. *In general, (7.5) holds for almost all* x.

<u>Proof</u>. We note that the conditions imposed on H imply that $H(t) \ge 0$ for $0 \le t < \infty$, and that

$$tH(t) \to 0, \text{ as } t \to +\infty, \text{ or } t \downarrow 0, \tag{7.7}$$

since

$$\frac{1}{2}\ tH(t) \le \int_{t/2}^{t} H(x)dx \to 0, \text{ as } t \to +\infty, \text{ or } t \downarrow 0,$$

so that there exists a constant C, such that $tH(t) \leq C$, for $0 \leq t < \infty$.

If we define

$$G(t) = \int_0^t g_x(y)dy, \quad t \geq 0, \tag{7.8}$$

then G is absolutely continuous. Because of assumption (7.6), given $\varepsilon > 0$, we can choose $\eta > 0$, such that $|G(t)| < \varepsilon t$, for $0 \leq t \leq \eta$. Having chosen such an η, we keep it fixed.

By Theorem 8 we have

$$\frac{1}{2\pi}(f*H_R)(x) - f(x) = \frac{1}{\pi}\int_0^\infty g_x(t)RH(Rt)dt = \frac{1}{\pi}\left(\int_0^\eta + \int_\eta^\infty\right) = I_1+I_2, \tag{7.9}$$

say. Then, for $R > 0$, we have, by (7.8),

$$I_1 = \frac{1}{\pi}\int_0^\eta RH(Rt)dG(t) = \frac{1}{\pi} RH(R\eta)G(\eta) + \frac{1}{\pi}\int_0^\eta G(t)\ R\ d\{-H(Rt)\},$$

by partial integration of the Stieltjes integral, where the integrator G is continuous and of bounded variation in $[0,\eta]$, and the integrand H is of bounded variation. By the choice of η, we have

$$|I_1| \leq \frac{\varepsilon}{\pi}\left[\eta RH(R\eta) + R\int_0^\eta t\ d\{-H(Rt)\}\right] = \frac{\varepsilon}{\pi}\left[\int_0^{\eta R} H(t)dt\right] < \varepsilon, \tag{7.10}$$

because of the particular normalization of H that has been assumed. As for I_2 in (7.9) we have

$$|I_2| = \left|\frac{1}{2\pi}\int_\eta^\infty [f(x+t) + f(x-t) - 2f(x)]RH(Rt)dt\right| \tag{7.11}$$

$$\leq \frac{1}{2\pi}\left[\frac{R\eta H(R\eta)}{\eta} 2\| f \|_1 + 2|f(x)| \int_{\eta R}^\infty H(t)dt\right] \to 0,$$

as $R \to \infty$, for a fixed $\eta > 0$. From (7.9), (7.10), and (7.11), we obtain (7.5). In any closed interval of points of continuity, f is *uniformly* continuous, so that the choice of $\eta = \eta(\varepsilon)$ in (7.9) can be made independently of x, and (7.5) then holds uniformly in that interval.

Finally we note that condition (7.6) is equivalent to the condition

$$\lim_{h\to 0} \frac{1}{2h}\int_{-h}^h f(x+t)dt = f(x),$$

which is satisfied for almost all x, since $f \in L_1(-\infty,\infty)$, because of Lebesgue's theorem that the indefinite integral of f is absolutely continuous and has a finite derivative, which equals f almost everywhere.

Remarks

1. We can take for $K(\alpha)$ the *Gauss kernel* $e^{-\alpha^2}$, or the *Abel kernel* $e^{-|\alpha|}$, and conclude that if $f \in L_1(-\infty,\infty)$, then at every point of continuity x of f, and for almost all x, we have

$$(7.12) \qquad \frac{1}{2\pi}\int_{-\infty}^{\infty} \hat{f}(\alpha)\, e^{-\alpha^2/R^2}\, e^{-i\alpha x} d\alpha \to f(x),$$

$$\frac{1}{2\pi}\int_{-\infty}^{\infty} \hat{f}(\alpha)\, e^{-|\alpha|/R}\, e^{-i\alpha x} d\alpha \to f(x),$$

as $R \to \infty$. An equivalent statement is the following

(7.13) *Corollary. If* $f \in L_1(-\infty,\infty)$, *then the Fourier integral*

$$\frac{1}{2\pi}\int_{-\infty}^{\infty} \hat{f}(\alpha)\, e^{-i\alpha x} d\alpha$$

is Gauss summable, and Abel summable, at every point of continuity x *of* f, *and for almost all* x, *to sum* f(x).

2. If $K(\alpha) = e^{-\alpha^2}$, then $\hat{K}(x) \equiv H(x) = \sqrt{\pi}\, e^{-x^2/4}$, and since $H_R(x) = RH(Rx)$, $R > 0$, we have

$$\frac{1}{2\pi}(f*H_R)(x) = \frac{1}{2\pi}\int_{-\infty}^{\infty} f(\xi)\, R\sqrt{\pi}\, e^{-R^2(x-\xi)^2/4} d\xi.$$

If we set $t = 1/R^2 > 0$, and

$$(7.14) \qquad W(x,t) = \frac{1}{2\sqrt{(\pi t)}}\, e^{-x^2/(4t)}, \quad t > 0,$$

the last integral becomes

$$(7.15) \qquad \int_{-\infty}^{\infty} f(\xi)\, W(x-\xi,t)\, d\xi = U(f;\, x,t),$$

say. This is referred to as the *Gauss-Weierstrass integral* of f, and because of (7.1) and (7.12) we have the following

(7.16) <u>*Corollary.*</u> *If* $f \in L_1(-\infty,\infty)$, *then the Gauss-Weierstrass integral* $U(f;\ x,t)$ *of* f, *given by (7.15), converges to* $f(x)$ *as* $t \to 0+$, *for almost all* x.

3. If $K(\alpha) = e^{-|\alpha|}$, then $\hat{K}(x) \equiv H(x) = \frac{2}{1+x^2}$, and on setting $t = 1/R > 0$, and

$$(7.17)\qquad P(x,t) = \frac{1}{\pi}\ \frac{t}{t^2+x^2},$$

we obtain the following

(7.18) <u>*Corollary.*</u> *The Cauchy-Poisson integral of* f, *namely*

$$V(f;\ x,t) = \int_{-\infty}^{\infty} f(\xi)\ P(x-\xi,t)d\xi,\ t > 0,$$

converges to $f(x)$ *as* $t \to 0+$, *for almost all* x.

<u>*Theorem 10*</u>. *Let* $f \in L_1(-\infty,\infty)$, *and for each* $x \in (-\infty,\infty)$ *let*

$$g_x(t) = \frac{1}{2}\,[f(x+t) + f(x-t) - 2f(x)],\ 0 \le t < \infty.$$

Let $K \in L_1(-\infty,\infty)$, K *even*, $\hat{K} = H$, *and* $\frac{1}{2\pi}\int_{-\infty}^{\infty} H(t)dt = 1$. *Suppose that there exists a function* H_0, *such that*

$$|H(t)| \le H_0(t) \in L_1(0 \le t < \infty),$$

and H_0 *is monotone decreasing in* $[0,\infty)$. *Then*

$$(7.5)\qquad \frac{1}{2\pi}\,(f*H_R)(x) \to f(x),\quad \text{as } R \to \infty,\quad \Big(H_R(x) = RH(Rx)\Big)$$

at every point x *at which*

$$(7.20)\qquad \lim_{h\to 0} \frac{1}{h}\int_0^h |g_x(t)|dt = 0;$$

in particular, at every point x *at which* f *is continuous; and uniformly over any closed interval of points of continuity of* f; *(7.5) holds for almost all* x, *in general.*

<u>Proof</u>. As in the proof of Theorem 9, given $\varepsilon > 0$, we choose η such that $|\tilde{G}(t)| < \varepsilon t$ for $0 \le t \le \eta$, where

$$\tilde{G}(t) = \int_0^t |g_x(u)|du ,$$

and write

$$\frac{1}{2\pi}(f*H_R)(x) - f(x) = \frac{1}{\pi}\left(\int_0^\eta + \int_\eta^\infty\right)g_x(t)RH(Rt)dt = I_1 + I_2, \text{ say.}$$

We then have

$$|I_1| \leq \int_0^\eta |g_x(t)|RH_0(Rt)dt \to 0, \quad \text{as } R\to\infty,$$

as in the proof of Theorem 9, while

$$|I_2| \leq \left|\int_\eta^\infty \frac{1}{2\pi}[f(x+t) + f(x-t) - 2f(x)]RH(Rt)dt\right|$$

$$\leq \frac{1}{2\pi}\,\frac{R\eta\ H_0(R\eta)}{\eta}\,2\|f\|_1 + \frac{|f(x)|}{\pi}\int_{\eta R}^\infty H_0(t)dt$$

$$\to 0, \text{ as } R\to\infty,$$

for a fixed $\eta > 0$, as before. By a theorem of Lebesgue, condition (7.20) holds almost everywhere for any function $f \in L_1(-\infty,\infty)$.

Remarks. We may take for $K(\alpha)$ in Theorem 10 the *Cesàro kernel* given by $K(\alpha) = 1 - |\alpha|$ for $|\alpha| \leq 1$, and $K(\alpha) = 0$ for $|\alpha| > 1$. Its Fourier transform $\hat{K}(t) \equiv H(t) = \left(\frac{\sin t/2}{t/2}\right)^2$ is the *Fejér Kernel* (cf. Example 2, §1), which is not monotone decreasing in $[0,\infty)$. But we may take $H_0(t) = \frac{c}{1+t^2}$, with a suitable constant $c > 0$, so that $|H(t)| \leq \frac{c}{1+t^2} = H_0(t)$, with $H_0(t) \in L_1(0 \leq t < \infty)$, and monotone decreasing in $[0,\infty)$. Thus we obtain the following analogue of Fejér's classical theorem on (trigonometric) Fourier series.

(7.21) *Corollary. If* $f \in L_1(-\infty,\infty)$. *then*

$$\lim_{R\to\infty} \frac{1}{2\pi}\int_{-R}^R \hat{f}(\alpha)\left(1 - \frac{|\alpha|}{R}\right)e^{-i\alpha x}d\alpha = f(x)$$

at every point of continuity of f, *and uniformly over any closed interval of points of continuity of* f, *and for almost all* x, *in general. The Fourier integral*

$$\frac{1}{2\pi}\int_{-\infty}^{\infty}\hat{f}(\alpha)\, e^{-i\alpha x}d\alpha$$

is, in other words, (C,1) *summable (Cesàro summable of order 1) to sum* f(x) *at every point of continuity of* f*, and uniformly in every closed interval of points of continuity, and for almost all* x *in general.*

§8. The inversion formula

The theorems on pointwise summability can be used to "invert" the Fourier transform almost everywhere.

Theorem 11. If $f\in L_1(-\infty,\infty)$*, and* $\hat{f}\in L_1(-\infty,\infty)$*, then we have*

$$f(x) = \frac{1}{2\pi}\int_{-\infty}^{\infty}\hat{f}(\alpha)\, e^{-i\alpha x}d\alpha$$

for almost all $x\in(-\infty,\infty)$.

Proof. By Theorem 9, Corollary (7.13), we have, for *almost* all x,

$$\lim_{R\to\infty}\frac{1}{2\pi}\int_{-\infty}^{\infty}\hat{f}(\alpha)\, e^{-|\alpha|/R}\, e^{-i\alpha x}d\alpha = f(x).$$

If $\hat{f}\in L_1(-\infty,\infty)$, then the left-hand side equals

$$\frac{1}{2\pi}\int_{-\infty}^{\infty}\hat{f}(\alpha)\, e^{-i\alpha x}d\alpha$$

by Lebesgue's theorem on dominated convergence.

Remarks. If $\hat{f}\in L_1(-\infty,\infty)$, the integral

$$\frac{1}{2\pi}\int_{-\infty}^{\infty}\hat{f}(\alpha)\, e^{-i\alpha x}d\alpha$$

defines a continuous function of x, so that the function $f\in L_1(-\infty,\infty)$ that we started with in Theorem 11 is continuous almost everywhere. Hence we obtain

Theorem 11'. If $f \in L_1(-\infty,\infty)$*, and* $\hat{f} \in L_1(-\infty,\infty)$*, and* f *is continuous in* $(-\infty,\infty)$*, then*

$$f(x) = \frac{1}{2\pi} \int_{-\infty}^{\infty} \hat{f}(\alpha)\, e^{-i\alpha x} d\alpha$$

for every $x \in (-\infty,\infty)$.

Examples

1. We have already (§1, Ex. 2) seen that if

$$K(x) = \begin{cases} 1-|x|, & |x| \le 1, \\ 0, & |x| > 1, \end{cases}$$

then $\hat{K}(\alpha) = \left(\frac{\sin \alpha/2}{\alpha/2}\right)^2$. Here both K and $\hat{K}$ belong to $L_1(-\infty,\infty)$. Hence we have, by Theorems 11 and 11',

$$(8.1) \qquad \frac{1}{2\pi} \int_{-\infty}^{\infty} \left(\frac{\sin \alpha/2}{\alpha/2}\right)^2 e^{-i\alpha x} d\alpha = \frac{1}{2\pi} \int_{-\infty}^{\infty} \left(\frac{\sin \alpha/2}{\alpha/2}\right)^2 e^{i\alpha x} d\alpha$$

$$= \begin{cases} 1-|x|, & |x| \le 1, \\ 0, & |x| > 1. \end{cases}$$

For $x = 0$, we get the formula

$$(8.2) \qquad \int_{-\infty}^{\infty} \frac{\sin^2 \alpha}{\alpha^2}\, d\alpha = \pi.$$

2. If $K(x) = e^{-a|x|}$, with $a > 0$, then $\hat{K}(\alpha) = \frac{2a}{a^2+\alpha^2}$ (Example 4, §1). By Theorems 11 and 11' we have

$$\frac{1}{2\pi} \int_{-\infty}^{\infty} \frac{2a}{a^2+\alpha^2}\, e^{-i\alpha x} d\alpha = \frac{2a}{\pi} \int_0^{\infty} \frac{\cos \alpha x}{a^2+\alpha^2}\, d\alpha = e^{-a|x|},$$

hence

$$(8.3) \qquad \int_0^{\infty} \frac{\cos \alpha x}{a^2+\alpha^2}\, d\alpha = \frac{\pi}{2a}\, e^{-a|x|}, \quad a > 0.$$

3. If $a > 0$, $b \ge 0$, we have the formula

$$(8.4) \qquad \int_0^{\infty} e^{-a^2x-b^2/x}\, x^{-1/2} dx = \frac{\sqrt{\pi}}{a}\, e^{-2ab}.$$

For

$$\Gamma(\mu+1) = \int_0^\infty e^{-x}\, x^\mu dx, \quad \text{for } \mu > -1,$$

so that

$$\frac{\Gamma(\mu+1)}{(x^2+a^2)^{\mu+1}} = \int_0^\infty e^{-(x^2+a^2)y}\, y^\mu dy, \quad (x \text{ real}, a > 0, \mu > -1)$$

hence

$$\int_0^\infty \frac{\cos \beta x}{(x^2+a^2)^{\mu+1}}\, dx = \frac{1}{\Gamma(\mu+1)} \int_0^\infty e^{-a^2 y}\, y^\mu dy \int_0^\infty e^{-x^2 y} \cos \beta x\, dx$$

$$= \frac{\sqrt{\pi}}{2\Gamma(\mu+1)} \int_0^\infty y^{\mu-\frac{1}{2}}\, e^{-a^2 y - \beta^2/4y} dy\ ,$$

if we use the expression for the Fourier transform of e^{-x^2} (Example 6, §1). On taking $\mu = 0$, $\beta = 2b \geq 0$, and using (8.3), we obtain (8.4).

The next theorem gives sufficient conditions for the Fourier transform of $f \in L_1(-\infty,\infty)$ to belong also to $L_1(-\infty,\infty)$.

Theorem 12. If $f(x) \in L_1(-\infty < x < \infty)$, *and there exists* $h > 0$, *such that* $|f(x)| \leq M < \infty$ *for* $-\infty < -h \leq x \leq h < +\infty$, *and* $\hat{f}(\alpha) \geq 0$ *for every* $\alpha \in (-\infty,\infty)$, *then we have*

$$\int_{-\infty}^{\infty} |\hat{f}(\alpha)|\, d\alpha = \int_{-\infty}^{\infty} \hat{f}(\alpha)\, d\alpha < \infty,$$

so that (by Theorem 11)

$$f(x) = \frac{1}{2\pi} \int_{-\infty}^{\infty} \hat{f}(\alpha)\, e^{-i\alpha x} d\alpha$$

for almost every $x \in (-\infty,\infty)$.

Proof. Let $K(\alpha) = e^{-|\alpha|}$, so that $\hat{K}(x) \equiv H(x) = \dfrac{2}{1+x^2}$, and $\frac{1}{2\pi}\int_{-\infty}^{\infty} H(t)dt = 1$. As in Theorem 8, (7.4), we have by the composition rule, for any $R > 0$,

$$\text{(8.5)} \qquad \int_{-\infty}^{\infty} \hat{f}(\alpha)\, e^{-|\alpha|/R}\, e^{-i\alpha x} d\alpha = \int_{-\infty}^{\infty} f(x+t)\, RH(Rt)\, dt$$

$$= \int_{-\infty}^{\infty} f(x+\tfrac{t}{R}) H(t)\, dt$$

For x = 0, we get

$$\int_{-\infty}^{\infty} \hat{f}(\alpha)\, e^{-|\alpha|/R} d\alpha = \int_{-\infty}^{\infty} f(\tfrac{t}{R}) H(t)\, dt \qquad , \; R > 0$$

$$= \left[\int_{-\infty}^{-hR} + \int_{-hR}^{hR} + \int_{hR}^{\infty}\right] f(\tfrac{t}{R}) H(t)\, dt = I_1 + I_2 + I_3,$$

say. We have

$$|I_2| \leq M \int_{-\infty}^{\infty} H(t)\, dt = M\cdot 2\pi, \text{ since } H(t) \geq 0.$$

Since tH(t) is bounded,

$$|I_3| \leq N_1 \int_{hR}^{\infty} |f(\tfrac{t}{R})|\, |t^{-1}|\, dt \leq \frac{N_1}{hR} \int_{hR}^{\infty} |f(\tfrac{t}{R})|\, dt \leq \frac{N_1}{h} \| f \|_1,$$

where N_1 is a constant, and similarly $|I_1| \leq \frac{N_2}{h} \| f \|_1$, where N_2 is a constant (since H is even). Hence

$$\int_{-\infty}^{\infty} \hat{f}(\alpha)\, e^{-|\alpha|/R} d\alpha < N < \infty,$$

where N is independent of R. Since $\hat{f}(\alpha) \geq 0$, we have, by Lebesgue's theorem on monotone convergence,

$$\lim_{R\to\infty} \int_{-\infty}^{\infty} \hat{f}(\alpha)\, e^{-|\alpha|/R} d\alpha = \int_{-\infty}^{\infty} \hat{f}(\alpha)\, d\alpha \leq N < \infty.$$

(8.6) *Corollary. If* $h > 0$, $f(x) \in L_1(-h \leq x \leq h)$, *and* $|f(x)| \leq M < \infty$, *for* $|x| \leq h$, *and*

$$\varphi(\alpha) \equiv \int_{-h}^{h} f(x)\, e^{i\alpha x} dx \geq 0,$$

then $\varphi(\alpha) \in L_1(-\infty < \alpha < \infty)$.

We have only to define $f(x) = 0$ for $|x| > h$, and use Theorem 12.

Remark. Instead of the kernel $e^{-|\alpha|}$, we could have used $e^{-\alpha^2}$ or the Cesàro kernel: $K(\alpha) = 1-|\alpha|$, for $|\alpha| \leq 1$, and $K(\alpha) = 0$ for $|\alpha| > 1$. Theorem 12 implies also the following

(8.7) *Corollary. If* $f \in L_1(-\infty,\infty)$, $\hat{f}(\alpha) \geq 0$ *for* $-\infty < \alpha < \infty$, *and* f *is continuous at the origin, then* $\hat{f} \in L_1(-\infty,\infty)$, *and*

(8.8) $$f(x) = \frac{1}{2\pi}\int_{-\infty}^{\infty} \hat{f}(\alpha)\, e^{-i\alpha x} d\alpha$$

for almost every $x \in (-\infty,\infty)$. *In particular,*

(8.9) $$f(0) = \frac{1}{2\pi}\int_{-\infty}^{\infty} \hat{f}(\alpha)\, d\alpha.$$

<u>Fourier transforms in $L_1(-\infty,\infty)\cdot\cap\cdot L_2(-\infty,\infty)$</u>

We shall denote by $L_1\cdot\cap\cdot L_2$ the class of functions f, such that $f \in L_1(-\infty,\infty)$ *and* $f \in L_2(-\infty,\infty)$.

(8.10) <u>*Lemma*</u>. *Let* $f,g \in L_1\cdot\cap\cdot L_2$. *Then the function* h *defined by*

(8.11) $$h(x) = \int_{-\infty}^{\infty} f(x+y)\overline{g(y)}dy = \int_{-\infty}^{\infty} f(x-y)\,\overline{g(-y)}dy$$

is bounded, and continuous, and belongs tc $L_1(-\infty,\infty)$.

<u>Proof</u>. If we define $G(y) = \overline{g(-y)}$, then $G \in L_1(-\infty,\infty)$, and h is the convolution $f*G$, hence $h \in L_1(-\infty,\infty)$, by Theorem 2. Since $|h(x)| \le \| f \|_2 \cdot \| g \|_2$, where $f,g \in L_2(-\infty,\infty)$, h is bounded. Since

$$|h(x+t) - h(x)| \le \left(\int_{-\infty}^{\infty} |f(x+t+y) - f(x+y)|^2 dy\right)^{1/2}\left(\int_{-\infty}^{\infty} |g(y)|^2 dy\right)^{1/2}$$

$$= \tau_f(t)\ \| g \|_2 \to 0 \text{ as } t \to 0, \quad \text{(cf. (1.17), §1)}$$

we see that h is continuous.

<u>*Theorem 13*</u>. *If* $f \in L_1\cdot\cap\cdot L_2$, *then* $\| f \|_2^2 = \frac{1}{2\pi}\| \hat{f} \|_2^2$, *so that the Fourier transform of* f *belongs to* $L_2(-\infty,\infty)$.

<u>Proof</u>. Let $F(x) = \overline{f(-x)}$, so that $F \in L_1\cdot\cap\cdot L_2$. Let $h = f*F \in L_1(-\infty,\infty)$. Since the Fourier transform of $\overline{f}(x)$ is $\overline{\hat{f}}(-\alpha)$, we see that $\hat{h}(\alpha) = |\hat{f}(\alpha)|^2 \ge 0$. Since h is bounded, and continuous, it follows from Theorem 12 that $\hat{h} \in L_1(-\infty,\infty)$, and that

$$h(x) = \int_{-\infty}^{\infty} f(x+y)\ \overline{f}(y)dy = \frac{1}{2\pi}\int_{-\infty}^{\infty} |\hat{f}(\alpha)|^2\, e^{-i\alpha x} d\alpha$$

for *every* $x \in (-\infty,\infty)$. On setting $x = 0$, we get the theorem.

Theorem 14. *If* $f, g \in L_1 \cap L_2$, *then* $\int_{-\infty}^{\infty} f(x)\overline{g(x)}dx = \frac{1}{2\pi}\int_{-\infty}^{\infty} \hat{f}(\alpha)\overline{\hat{g}}(\alpha)d\alpha$.

Proof. By Theorem 13 we have $\|\hat{f}\|_2 < \infty$, $\|\hat{g}\|_2 < \infty$, so that $\|\hat{f}\,\hat{g}\|_1 < \infty$ by Schwarz's inequality. If $h(x)$ is defined as in (8.11), $h = f*G$, where $G(y) = \overline{g(-y)}$, then $h \in L_1(-\infty,\infty)$, and $\hat{h}(\alpha) = \hat{f}(\alpha)\cdot\overline{\hat{g}}(\alpha)$, and since h is continuous, its Fourier transform can be inverted everywhere (Theorem 11'), so that

$$h(x) = \frac{1}{2\pi}\int_{-\infty}^{\infty} \hat{f}(\alpha)\,\overline{\hat{g}}(\alpha)\,e^{-i\alpha x}d\alpha,$$

and on setting $x = 0$ we get the required result.

Examples

1. If $f(x) = 1$, for $|x| \le 1$; and $f(x) = 0$, for $|x| > 1$, then $\hat{f}(\alpha) = 2\left(\frac{\sin\alpha}{\alpha}\right)$ (cf. Example 1, §1). Theorem 13 gives the formula

$$\int_{-\infty}^{\infty} \frac{\sin^2\alpha}{\alpha^2}\,d\alpha = \pi.$$

2. If $a > 0$, and $f(x) = 1 - \frac{|x|}{a}$, for $|x| \le a$; and $f(x) = 0$, for $|x| > a$, then $\hat{f}(\alpha) = \dfrac{\sin^2(\frac{\alpha a}{2})}{a(\alpha/2)^2}$ (cf. Example 2, §1), and Theorem 13 gives the formula

$$\int_{-\infty}^{\infty} \left(\frac{\sin b\alpha}{\alpha}\right)^4 d\alpha = \frac{2\pi}{3}\,b^3, \quad \text{for any } b \ge 0.$$

3. If $a > 0$, and $f(x) = e^{-a|x|}$, then $\hat{f}(\alpha) = \frac{2a}{a^2+\alpha^2}$ (cf. Example 4, §1). For $b > 0$, let $g(x) = e^{-b|x|}$. Then Theorem 14 gives the formula

$$\int_0^{\infty} \frac{d\alpha}{(a^2+\alpha^2)(b^2+\alpha^2)} = \frac{\pi}{2ab(a+b)}$$

4. If $f(x) = 1$, for $|x| \le a$; and $f(x) = 0$, for $|x| > a$, where $a > 0$, and $g(x) = 1$, for $|x| \le b$; and $g(x) = 0$, for $|x| > b$, where $b > 0$, then Theorem 14 gives the formula

$$\int_0^{\infty} \frac{\sin ax\cdot\sin bx}{x^2}\,dx = \frac{\pi}{2}\,\min(a,b).$$

Fourier transforms in S

We have considered in §3 Schwartz's space of infinitely differentiable functions f which are "rapidly decreasing". We have noted in (3.7) that *if* $f \in S$, *then* $\hat{f} \in S$, S being a *dense* subspace of $L_p(-\infty,\infty)$ for every p, $1 \leq p < \infty$. It is a trivial consequence of Theorem 11' that if $f \in S$, then

$$f(x) = \frac{1}{2\pi} \int_{-\infty}^{\infty} \hat{f}(\alpha)\, e^{-i\alpha x} d\alpha \tag{8.12}$$

for every $x \in (-\infty,\infty)$, so that the *Fourier transform maps* S *onto itself*. Further we have for any two functions $f, g \in S$,

$$\int_{-\infty}^{\infty} f(x)\overline{g}(x)\,dx = \frac{1}{2\pi} \int_{-\infty}^{\infty} \hat{f}(\alpha)\overline{\hat{g}}(\alpha)\,d\alpha. \tag{8.13}$$

For $\hat{\overline{g}} \in S$, and since $g(x) = \frac{1}{2\pi} \int_{-\infty}^{\infty} \hat{g}(\alpha)\, e^{-i\alpha x} d\alpha$, we obtain $\overline{g}(x) = \frac{1}{2\pi} \int_{-\infty}^{\infty} \overline{\hat{g}}(\alpha)\, e^{i\alpha x} d\alpha = \frac{1}{2\pi}\, \hat{\overline{\hat{g}}}(x)$. By the composition rule (1.13), however, we have

$$\int_{-\infty}^{\infty} f(x)\, \hat{\overline{\hat{g}}}(x)\,dx = \int_{-\infty}^{\infty} \hat{f}(\alpha)\, \overline{\hat{g}}(\alpha)\,d\alpha,$$

or

$$\int_{-\infty}^{\infty} f(x)\, 2\pi\overline{g}(x)\,dx = \int_{-\infty}^{\infty} \hat{f}(\alpha)\, \overline{\hat{g}}(\alpha)\,d\alpha,$$

as claimed. On taking $g = f$, we get

$$\int_{-\infty}^{\infty} |f(x)|^2 dx = \frac{1}{2\pi} \int_{-\infty}^{\infty} |\hat{f}(\alpha)|^2 d\alpha .$$

These are special cases of Theorem 14, but simpler to prove directly.

Finally, if $f, g \in S$, then $\hat{f}, \hat{g} \in S$, hence also $\hat{f}\cdot\hat{g} \in S$. But $(f*g)^\wedge = \hat{f}\cdot\hat{g}$, by Theorem 2. Hence

$$\frac{1}{2\pi} \int_{-\infty}^{\infty} \hat{f}(\alpha)\, \hat{g}(\alpha)\, e^{-i\alpha x} dx \in S, \tag{8.14}$$

which means that $f*g \in S$.

In the notation of (1.2) we can write (8.13) as

$$\int_{-\infty}^{\infty} f(x)\, \overline{g}(x)\, dx = \int_{-\infty}^{\infty} F[f](\alpha)\, \overline{F[g]}(\alpha)\,d\alpha. \tag{8.15}$$

The fact that S is a dense subset of $L_2(-\infty,\infty)$ leads (in Ch.II) to Plancherel's theorem.

§9. Summability in the L_1-norm

We have seen in §7 that if $f \in L_1(-\infty,\infty)$, for special choices of the kernel K, we have

$$\lim_{R\to\infty} \frac{1}{2\pi} \int_{-\infty}^{\infty} \hat{f}(\alpha)\ K(\tfrac{\alpha}{R})\ e^{-i\alpha x} d\alpha = f(x),$$

pointwise almost everywhere. It is somewhat simpler to consider this limit in the L_1-norm. Before doing so, we shall prove a general result on approximating any $f \in L_1(-\infty,\infty)$ in the L_1-norm.

Theorem 15. *Let* $H \in L_1(-\infty,\infty)$, *with* $\frac{1}{2\pi}\int_{-\infty}^{\infty} H(\alpha)\ d\alpha = 1$, *and for* $R > 0$ *let* $H_R(\alpha) = R\ H(R\alpha)$.

If $f \in L_1(-\infty,\infty)$, then we have

$$\| \frac{1}{2\pi} (f*H_R) - f\|_1 \to 0, \quad \text{as } R\to\infty. \tag{9.1}$$

Proof. By definition we have

$$\frac{1}{2\pi} (f*H_R)(x) = \frac{1}{2\pi} \int_{-\infty}^{\infty} f(x-y)\ RH(Ry)dy\ ,$$

and, by assumption, we have

$$\frac{1}{2\pi} \int_{-\infty}^{\infty} H_R(y)dy = \frac{1}{2\pi} \int_{-\infty}^{\infty} RH(Ry)dy = \frac{1}{2\pi} \int_{-\infty}^{\infty} H(\alpha)d\alpha = 1.$$

Hence

$$\frac{1}{2\pi} (f*H_R)(x) - f(x) = \frac{1}{2\pi} \int_{-\infty}^{\infty} [f(x-y) - f(x)]\ RH(Ry)dy\ ,$$

and

$$\| \frac{1}{2\pi} (f*H_R) - f\|_1 \le \frac{1}{2\pi} \int_{-\infty}^{\infty} dx \int_{-\infty}^{\infty} |f(x-y) - f(x)|\cdot R|H(Ry)|dy$$

$$= \frac{1}{2\pi} \int_{-\infty}^{\infty} \tau_f(y)\ R|H(Ry)|dy \qquad \text{(cf. (1.14))}$$

where $\tau_f(y)$ is the L_1-modulus of continuity of f (see (1.14)), which is *bounded*, even, non-negative, and tends to zero as $y \to 0$. Given

$\varepsilon > 0$, we can choose $\eta > 0$, such that $0 \leq \tau_f(y) < \dfrac{\varepsilon \cdot 2\pi}{\|H\|_1}$ for $|y| \leq \eta$. We then write

$$\left\| \frac{1}{2\pi} (f*H_R) - f \right\|_1 \leq \frac{1}{2\pi} \left(\int_{|y|\leq\eta} + \int_{|y|>\eta} \right) \tau_f(y)\ R|H(Ry)|\,dy$$

$$= I_1 + I_2, \text{ say,}$$

where

$$|I_2| \leq c \int_{|t|>\eta R}^{\infty} |H(t)|\,dt \to 0, \text{ as } R \to \infty,$$

(c being a suitable constant), while

$$|I_1| \leq \frac{1}{2\pi} \sup_{|y|\leq\eta} \tau_f(y)\ \|H\|_1 < \varepsilon,$$

by the choice of η, and the theorem follows.

Remarks

If $H(x) = \left(\dfrac{\sin x/2}{x/2}\right)^2$, then $H \in L_1(-\infty,\infty)$, and $\dfrac{1}{2\pi} \int_{-\infty}^{\infty} H(\alpha)\,d\alpha = 1$, (see (8.1)), and

$$\frac{1}{2\pi} \hat{H}(\alpha) = \frac{1}{2\pi} \int_{-\infty}^{\infty} H(x)\ e^{i\alpha x} dx = \begin{cases} 1 - |\alpha|, & \text{for } |\alpha| \leq 1, \\ 0, & \text{for } |\alpha| > 1; \end{cases}$$

$$= K(\alpha), \text{ say.}$$

Then

$$\frac{1}{2\pi} \int_{-\infty}^{\infty} H_R(y)\ e^{i\alpha y} dy = \frac{1}{2\pi} \int_{-\infty}^{\infty} R\ H(Ry)\ e^{i\alpha y} dy$$

$$= \frac{1}{2\pi} \int_{-\infty}^{\infty} H(x)\ e^{i\alpha x/R} dx = K\left(\frac{\alpha}{R}\right) \equiv K_R(\alpha), \text{ say,}$$

where K_R vanishes outside the interval $[-R,R]$. The Fourier transform of $(1/2\pi)\ (f*H_R)$, in Theorem 15, is $\hat{f} \cdot K_R$, which vanishes therefore outside $[-R,R]$, and from Theorem 15 we can deduce the following

(9.2) *Corollary. Every function* $f \in L_1(-\infty,\infty)$ *can be approximated in the* L_1*-norm by a function in* $L_1(-\infty,\infty)$ *whose Fourier transform* *vanishes outside a bounded interval.*

Theorem 15 has also another interpretation. There is no *unit element* relative to multiplication in the L_1-algebra over $(-\infty,\infty)$. That is to say, there exists no function $I \in L_1(-\infty,\infty)$ such that $I*f = f$, for *every* $f \in L_1(-\infty,\infty)$. For if it did, we would have, in particular, $I*I = I$, which implies that $\hat{I}(\alpha) = \{\hat{I}(\alpha)\}^2$, hence $\hat{I}(\alpha) = 0$, or 1, for each given α. Since $\hat{I}(\alpha)$ is continuous, we must have $\hat{I}(\alpha) \equiv 0$ or $\hat{I}(\alpha) \equiv 1$. By the Riemann-Lebesgue theorem, however, $\hat{I}(\alpha) \to 0$ as $|\alpha| \to \infty$. Hence we must have $\hat{I}(\alpha) \equiv 0$ identically. By the uniqueness theorem for the Fourier transform (Theorem 7), it follows that $I(\alpha) = 0$ for almost all α. If $f \in L_1(-\infty,\infty)$ is such that it is non-zero almost everywhere, the equation $I*f = f$ will be contradicted. We have, however, an *approximate unit* in $L_1(-\infty,\infty)$, by which we mean that we can find a sequence of functions (δ_n), such that $\delta_n \geq 0$, $\delta_n \in L_1(-\infty,\infty)$, $\|\delta_n\|_1 = 1$, for each n, and such that $\delta_n*f \to f$ in the L_1-norm for *every* $f \in L_1(-\infty,\infty)$. Theorem 15, and Corollary (9.2), show that $H_n(x) = \frac{1}{2\pi}\,\frac{(\sin \frac{nx}{2})^2}{n(x/2)^2}$ is such an approximate unit.

<u>*Theorem 16*</u>. *Let* $K \in L_1(-\infty,\infty)$, K *even*, $\hat{K} \equiv H \in L_1(-\infty,\infty)$, $\frac{1}{2\pi}\int_{-\infty}^{\infty} H(\alpha)d\alpha = 1$. *Let* $R > 0$, *and* $H_R(\alpha) = R\,H(R\alpha)$.

If $f \in L_1(-\infty,\infty)$, then

$$(9.3) \qquad \left\| \frac{1}{2\pi}(f*H_R) - f \right\|_1 \to 0, \text{ as } R \to \infty,$$

where (as in (7.1))

$$(9.4) \qquad \frac{1}{2\pi}(f*H_R)(x) = \frac{1}{2\pi}\int_{-\infty}^{\infty} \hat{f}(\alpha)\, K\left(\frac{\alpha}{R}\right) e^{-i\alpha x} d\alpha.$$

<u>Proof.</u> Since $f, H \in L_1(-\infty,\infty)$, we note that $f*H_R \in L_1(-\infty,\infty)$, and the integral equalling the convolution in (9.4) exists for *every* x, since $K \in L_1(-\infty,\infty)$ and $\hat{f}$ is bounded. To prove the theorem we have only to use Theorem 15.

<u>Remarks</u>

By taking $K(\alpha) = e^{-\alpha^2}$, and making use of (7.14) and (7.15), we deduce the following

(9.5) *Corollary. If* $f \in L_1(-\infty,\infty)$, *then the Gauss-Weierstrass integral*

$$U(f;x,t) = \int_{-\infty}^{\infty} f(\xi)\, W(x-\xi,t)\,d\xi, \quad W(x,t) = \frac{1}{2\sqrt{(\pi t)}}\, e^{-x^2/4t}, \quad t>0,$$

of f *converges in the* L_1*-norm to* f(x), *as* $t \to 0+$.

By taking $K(\alpha) = e^{-|\alpha|}$, and making use of (7.17), (7.19), we deduce the following

(9.6) *Corollary. If* $f \in L_1(-\infty,\infty)$, *then the Cauchy-Poisson integral*

$$V(f;x,t) = \int_{-\infty}^{\infty} f(\xi)\, P(x-\xi,t)\,d\xi, \quad P(x,t) = \frac{1}{\pi}\,\frac{t}{t^2+x^2}, \quad t>0,$$

of f *converges in the* L_1*-norm to* f(x), *as* $t \to 0+$.

The principal deduction from Theorem 16, which results from taking for $K(\alpha)$ the Abel, Gauss, and Cesàro kernels separately, as in (7.12) and (7.21), is

(9.7) *Corollary. The 'Fourier integral'*

$$\frac{1}{2\pi}\int_{-\infty}^{\infty} \hat{f}(\alpha)\, e^{-i\alpha x} d\alpha, \quad f \in L_1(-\infty,\infty)$$

is Abel, Gauss, and Cesàro (C,1) *summable in the* L_1*-norm to* f(x).

This is just another way of expressing (9.3) and (9.4).

For instance, in the case of the Gauss kernel, we have:

$$\frac{1}{2\pi}\int_{-\infty}^{\infty} \hat{f}(\alpha)\, e^{-\alpha^2/R^2}\, e^{-i\alpha x} d\alpha \to f(x), \text{ as } R \to \infty, \quad \text{in the } L_1\text{-norm.}$$

By Weyl's formulation of the Riesz-Fischer theorem, there exists a sequence $(R_k) \to \infty$ as $k \to \infty$, such that

$$\frac{1}{2\pi}\int_{-\infty}^{\infty} \hat{f}(\alpha)\, e^{-\alpha^2/R_k^2}\, e^{-i\alpha x} d\alpha \to f(x), \quad \text{as } k \to \infty,$$

for *almost every* $x \in (-\infty,\infty)$. If we *assume*, in addition, that $\hat{f} \in L_1(-\infty,\infty)$, then, by Lebesgue's theorem on dominated convergence, we obtain Theorem 11 on Fourier inversion.

As another application of Theorem 16 we shall prove

__Theorem 17__. *Let* $f,g \in L_1(-\infty,\infty)$. *If* $\hat{g}(\alpha) = -i\alpha\hat{f}(\alpha)$, *then we have*

$$f(x) = -\int_x^\infty g(y)dy .$$

__Proof.__ __Case (i)__. Let us assume, in addition, that $\hat{g},\hat{f} \in L_1(-\infty,\infty)$. Then, by Theorem 11, we have

$$f(x) = \frac{1}{2\pi}\int_{-\infty}^{\infty} \hat{f}(\alpha)\, e^{-i\alpha x}d\alpha$$

and

$$g(x) = \frac{1}{2\pi}\int_{-\infty}^{\infty} \hat{g}(\alpha)\, e^{-i\alpha x}d\alpha = \frac{1}{2\pi}\int_{-\infty}^{\infty} (-i\alpha)\hat{f}(\alpha)\, e^{-i\alpha x}d\alpha,$$

for *almost all* x, so that

$$\int_a^b g(x)dx = \frac{1}{2\pi}\int_{-\infty}^{\infty} \left(e^{-ib\alpha} - e^{-ia\alpha}\right) \hat{f}(\alpha)d\alpha, \quad -\infty < a < b < +\infty,$$

$$= f(b) - f(a),$$

hence $g(x) = f'(x)$ for *almost all* x. (Note that if $G(x) = \int_0^x g(y)dy$, where $g \in L_1(-\infty,\infty)$, then $G'(x) = g(x)$ for almost all x. And if $G(b) - G(a) = f(b) - f(a)$, for all a,b, such that $-\infty < a < b < +\infty$, then G differs from f by a constant, so that $f' = G' = g$ almost everywhere.)

__Case (ii)__. Let $K(\alpha) = e^{-\alpha^2}$, $-\infty < \alpha < \infty$, and let $\hat{K}(\alpha) \equiv H(\alpha) = \sqrt{\pi}\, e^{-\alpha^2/4}$. For $R > 0$, define

$$(9.8)\quad F_R(x) = \frac{1}{2\pi}\int_{-\infty}^{\infty} e^{-i\alpha x}\, \hat{f}(\alpha)\, K(\tfrac{\alpha}{R})d\alpha,$$

and

$$G_R(x) = \frac{1}{2\pi}\int_{-\infty}^{\infty} e^{-i\alpha x}\, \hat{g}(\alpha)\, K(\tfrac{\alpha}{R})d\alpha = \frac{1}{2\pi}\int_{-\infty}^{\infty} e^{-i\alpha x}(-i\alpha\hat{f}(\alpha))K(\tfrac{\alpha}{R})d\alpha.$$

We note that $\hat{f}$ is bounded, and that F_R is the Fourier transform of a function in $L_1(-\infty,\infty)$, and therefore $F_R(x) \to 0$ as $|x| \to \infty$.

By the composition rule (1.13), we have

$$F_R(x) = \frac{1}{2\pi} \int_{-\infty}^{\infty} f(x+t)\ RH(Rt)dt ,$$

so that

$$(9.9) \qquad \int_{-\infty}^{\infty} |F_R(x)|dx \le \frac{1}{2\pi} \|f\|_1 \int_{-\infty}^{\infty} RH(Rt)dt = \|f\|_1 < \infty$$

because of the choice of K. Hence $F_R \in L_1(-\infty,\infty)$, and similarly also $G_R \in L_1(-\infty,\infty)$, for each $R > 0$.

Since

$$F_R(b) - F_R(a) = \int_a^b G_R(x)dx, \qquad b > a,$$

we have, on letting $b \to \infty$,

$$F_R(x) = - \int_x^{\infty} G_R(y)dy.$$

Since $g \in L_1(-\infty,\infty)$, we have, by Theorem 16,

$$\lim_{R\to\infty} \int_{-\infty}^{\infty} |G_R(y) - g(y)|dy = 0,$$

which implies that

$$\lim_{R\to\infty} \int_{-\infty}^{\infty} G_R(y)dy = \int_{-\infty}^{\infty} g(y)dy .$$

Hence, for every fixed x, we have

$$\lim_{R\to\infty} F_R(x) = \lim_{R\to\infty} \left(- \int_x^{\infty} G_R(y)dy\right) = - \int_x^{\infty} g(y)dy .$$

On the other hand, by Theorem 9, $F_R(x) \to f(x)$, for *almost all* x, as $R \to \infty$. Hence

$$f(x) = - \int_x^{\infty} g(y)dy ,$$

for almost all $x \in (-\infty,\infty)$.

§10. The central limit theorem

As an illustration of the method of Fourier transforms, we shall state and prove a theorem which corresponds to what is known as the central limit theorem in the theory of probability.

Theorem 18. Let $f \in L_1(-\infty,\infty)$, $f(x) \geq 0$, $\int_{-\infty}^{\infty} f(x)dx = 1$, $\int_{-\infty}^{\infty} xf(x)dx = 0$, *and* $\int_{-\infty}^{\infty} x^2 f(x)dx = 1$, *and let* $f^n = f*\ldots*f$, *the convolution of* f *with itself* n *times. Then we have*

$$\lim_{n\to\infty} \int_{a\sqrt{n}}^{b\sqrt{n}} f^n(x)dx = \int_a^b \frac{e^{-x^2/2}}{\sqrt{(2\pi)}}dx, \quad -\infty < a < b < \infty.$$

Proof. Let

$$\chi_{a,b}(x) = \begin{cases} 1, & \text{for } a \leq x \leq b, \\ 0, & \text{otherwise,} \end{cases}$$

so that

$$\int_{-\infty}^{\infty} \sqrt{n}\, f^n(x\sqrt{n})\chi_{a,b}(x)dx = \int_a^b \sqrt{n}\, f^n(x\sqrt{n})dx = \int_{a\sqrt{n}}^{b\sqrt{n}} f^n(y)dy$$

and

$$\int_{-\infty}^{\infty} \frac{e^{-x^2/2}}{\sqrt{(2\pi)}} \chi_{a,b}(x)dx = \int_a^b \frac{e^{-x^2/2}}{\sqrt{(2\pi)}} dx.$$

It is sufficient therefore to prove that

$$(10.1) \qquad \lim_{n\to\infty} \int_{-\infty}^{\infty} \sqrt{n}\, f^n(x\sqrt{n})\ \chi_{a,b}(x)dx = \int_{-\infty}^{\infty} \frac{e^{-x^2/2}}{\sqrt{(2\pi)}} \chi_{a,b}(x)dx\ .$$

In order to prove this, it is sufficient, in turn, to prove that

$$(10.2) \qquad \lim_{n\to\infty} \int_{-\infty}^{\infty} \sqrt{n}\, f^n(x\sqrt{n})\ k(x)dx = \int_{-\infty}^{\infty} \frac{e^{-x^2/2}}{\sqrt{(2\pi)}} k(x)dx\ ,$$

where $k \in S$, where S denotes Schwartz's space of infinitely differentiable functions on $(-\infty,\infty)$ which are rapidly decreasing, defined in §3.

For, given $\chi_{a,b}$ we can find two functions $k_1, k_2 \in \mathcal{D} \subset S$, where $\mathcal{D}$ denotes the subspace of infinitely differentiable functions on $(-\infty,\infty)$ with bounded supports, such that for any $\varepsilon > 0$, we have

$$(10.3) \qquad k_1(x) < \chi_{a,b}(x) < k_2(x),\ \underline{\text{and}}\ \int_{-\infty}^{\infty} [k_2(x) - k_1(x)]dx < 4\varepsilon.$$

We have only to define (as in (3.12), (3.13))

$$k_1(x) = \begin{cases} 1, & a+\varepsilon \le x \le b-\varepsilon \\ 0, & x \le a,\ x \ge b \end{cases} \qquad k_2(x) = \begin{cases} 1, & a \le x \le b \\ 0, & x \le a-\varepsilon,\ x > b+\varepsilon \end{cases}$$

both k_1 and k_2 being infinitely differentiable everywhere.

If (10.2) is proved, then it holds with $k = k_1$ and $k = k_2$. Since

$$\int_{-\infty}^{\infty} \sqrt{n}\, f^n(\sqrt{n}\, x)\chi_{a,b}(x)dx - \int_{-\infty}^{\infty} \frac{e^{-x^2/2}}{\sqrt{(2\pi)}} \chi_{a,b}(x)dx$$

$$< \int_{-\infty}^{\infty} \sqrt{n}\, f^n(x\sqrt{n})k_2(x)dx - \int_{-\infty}^{\infty} \frac{e^{-x^2/2}}{\sqrt{(2\pi)}} k_2(x)dx + \int_{-\infty}^{\infty} \frac{e^{-x^2/2}}{\sqrt{(2\pi)}} k_2(x)dx - \int_{-\infty}^{\infty} \frac{e^{-x^2/2}}{\sqrt{(2\pi)}} \chi_{a,b}(x)dx$$

$$< \varepsilon + 4\varepsilon, \quad \text{for } n \ge n_0,$$

by (10.3), we obtain

$$\int_{-\infty}^{\infty} \sqrt{n}\, f^n(x\sqrt{n})\chi_{a,b}(x)dx - \int_{-\infty}^{\infty} \frac{e^{-x^2/2}}{\sqrt{(2\pi)}} \chi_{a,b}(x)dx < 5\varepsilon, \quad \text{for } n \ge n_0;$$

and by using k_1 in place of k_2, we see that the left-hand side is greater than -5ε for $n \ge n'$, thus proving (10.1).

It remains to prove (10.2). If $k \in S$, we have seen that $\hat{k} \in S$, and by Theorem 11', inversion holds everywhere. Hence

$$\int_{-\infty}^{\infty} \sqrt{n}\, f^n(x\sqrt{n})k(x)dx = \int_{-\infty}^{\infty} \sqrt{n}\, f^n(x\sqrt{n})\left\{\frac{1}{2\pi} \int_{-\infty}^{\infty} \hat{k}(\alpha)e^{-i\alpha x}d\alpha\right\}dx$$

$$= \frac{1}{2\pi} \int_{-\infty}^{\infty} \hat{k}(\alpha) \left(\int_{-\infty}^{\infty} \sqrt{n}\, f^n(x\sqrt{n})\ e^{-i\alpha x}dx\right)d\alpha$$

$$= \frac{1}{2\pi} \int_{-\infty}^{\infty} \hat{k}(\alpha) \left(\int_{-\infty}^{\infty} f^n(y)e^{-i\alpha y/\sqrt{n}}dy\right)d\alpha$$

$$(10.4) \qquad = \frac{1}{2\pi} \int_{-\infty}^{\infty} \hat{k}(\alpha) \left\{\hat{f}(-\tfrac{\alpha}{\sqrt{n}})\right\}^n d\alpha\ , \quad \text{(Theorem 2)}$$

where $\hat{f}(-\frac{\alpha}{\sqrt{n}})$ denotes the Fourier transform of f at the point $\frac{-\alpha}{\sqrt{n}}$.
Now

$$(10.5)\qquad \left|\hat{f}(-\tfrac{\alpha}{\sqrt{n}})\right|^n \le \left(\|f\|_1\right)^n = 1, \quad -\infty < \alpha < \infty,\ \hat{f}(0) = 1,$$

and for every fixed α, $\alpha \neq 0$, $-\infty < \alpha < \infty$, we have

$$(10.6)\qquad \hat{f}\left(-\frac{\alpha}{\sqrt{n}}\right) = \int_{-\infty}^{\infty} f(x)e^{-(i\alpha x)/\sqrt{n}}dx$$

$$= \int_{-\infty}^{\infty} f(x)\left\{1 - \frac{i\alpha x}{\sqrt{n}} - \frac{\alpha^2x^2}{2n}(1+r(n,x))\right\}dx \quad ,$$

where $r(n,x)$ is bounded uniformly in x and n, and $r(n,x) \to 0$, as $n \to \infty$, for every fixed x, $-\infty < x < \infty$. (Note that if $|\frac{x}{\sqrt{n}}| \le 1$, then $|r(n,x)| < c$, where c is a constant independent of x and n, while $|\frac{x}{\sqrt{n}}| > 1$ implies that

$$|1 + r(n,x)| = \left|\left(-e^{-i\alpha x/\sqrt{n}} + 1 - \frac{i\alpha x}{\sqrt{n}}\right)\left(\frac{2n}{\alpha^2x^2}\right)\right| = O(1),$$

the constant implied by the $O(1)$ being independent of x and n). It follows that

$$\lim_{n\to\infty}\int_{-\infty}^{\infty} x^2r(n,x)f(x)dx = \int_{-\infty}^{\infty} x^2f(x)\left\{\lim_{n\to\infty} r(n,x)\right\}dx = o(1),$$

as $n \to \infty$, since $\int_{-\infty}^{\infty} x^2f(x)dx = 1$, and $f(x) \ge 0$. By hypothesis, we also have

$$\int_{-\infty}^{\infty} f(x)dx = 1, \int_{-\infty}^{\infty} xf(x)dx = 0,$$

so that (10.6) yields the relation

$$(10.7)\qquad \hat{f}(-\frac{\alpha}{\sqrt{n}}) = 1 - \frac{\alpha^2}{2n}(1 + o(1)), \quad \text{as } n \to \infty,$$

which holds also for $\alpha = 0$, since $\hat{f}(0) = 1$. Hence

$$(10.8)\qquad \left\{\hat{f}(-\frac{\alpha}{\sqrt{n}})\right\}^n = \left\{1 - \frac{(\alpha/\sqrt{2})^2}{n}(1 + o(1))\right\}^n \to e^{-(\alpha/\sqrt{2})^2}, \text{ as } n \to \infty,$$

since $\lim_{n\to\infty}\left(1 - \frac{x}{n}\right)^n = e^{-x}$. If we take (10.4), and use (10.5), then by Lebesgue's theorem on dominated convergence, we obtain

$$\lim_{n\to\infty}\int_{-\infty}^{\infty} \sqrt{n}\, f^n(x\sqrt{n})\, k(x)dx = \frac{1}{2\pi}\int_{-\infty}^{\infty} \hat{k}(\alpha) \lim_{n\to\infty}\left\{\hat{f}(-\frac{\alpha}{\sqrt{n}})\right\}^n d\alpha$$

$$= \frac{1}{2\pi} \int_{-\infty}^{\infty} \hat{k}(\alpha)\ e^{-\alpha^2/2} d\alpha, \quad \text{by (10.8)}$$

$$= \frac{1}{2\pi} \int_{-\infty}^{\infty} k(\alpha)\ \sqrt{(2\pi)}\ e^{-\alpha^2/2} d\alpha,$$

if we use the composition rule (1.13) and the Fourier transform of $e^{-\alpha^2/2}$ (see §1, Example 6). Thus (10.2) is proved, which, as we have already shown, implies (10.1) and the theorem.

§11. Analytic functions of Fourier transforms

If $\hat{f}(x) = 1$ for all x, $-\infty < x < \infty$, then obviously (by Theorem 1) $\hat{f}$ cannot be the Fourier transform of a function in $L_1(-\infty,\infty)$. If, instead of the interval $(-\infty,\infty)$, we had only a bounded interval, say [a,b], then there exists a function $f \in L_1(-\infty,\infty)$, such that its Fourier transform $\hat{f}(\alpha) = 1$ for $\alpha \in [a,b]$, and $\hat{f}$ vanishes outside a larger interval.

We have constructed in (3.12) an *infinitely differentiable* function ω, which equals 1 in [a,b], and vanishes outside $(a-\varepsilon,\ b+\varepsilon)$, where $\varepsilon > 0$. Such a function belongs to Schwartz's space S, which has the property that if $f \in S$, then $\hat{f} \in S$ (see (3.7)). Further, as a trivial consequence of the inversion formula (Theorem 11'), and the fact that $S \subset L_1(-\infty,\infty)$, we note that if we *define* $\check{\omega}$ by the relation

$$\check{\omega}(x) = \frac{1}{2\pi} \int_{-\infty}^{\infty} \omega(t) e^{-itx} dt, \quad -\infty < x < \infty,$$

then $\check{\omega} \in S \subset L_1(-\infty,\infty)$, and

$$\omega(\alpha) = \int_{-\infty}^{\infty} \check{\omega}(x) e^{i\alpha x} dx \quad .$$

Hence we can assert the following (by taking $\check{\omega} = \delta$):

(11.1) Given two real numbers a,b with $b > a$, and a number $\varepsilon > 0$, there exists a function $\delta \in L_1(-\infty,\infty)$, such that its Fourier transform $\hat{\delta}$ has the property

$$\hat{\delta}(\alpha) = \begin{cases} 1, & a \leq \alpha \leq b, \\ 0, & \alpha \leq a-\varepsilon,\ \alpha \geq b+\varepsilon,\ \varepsilon > 0, \\ \text{infinitely differentiable in } (-\infty < \alpha < \infty). \end{cases}$$

(11.2) There exists a function $f \in L_1(-\infty,\infty)$, such that its Fourier transform $\hat{f}$ has the property

$$\hat{f}(\alpha) > 0, \quad \text{for } \alpha > 0,$$

and

$$\hat{f}(\alpha) = 0, \quad \text{for } \alpha \leq 0.$$

For if

$$F(x) = \begin{cases} x\, e^{-x}, & \text{for } x > 0, \\ 0, & \text{for } x \leq 0, \end{cases}$$

then $F \in L_1(-\infty,\infty)$, and

$$\hat{F}(-x) = \int_{-\infty}^{\infty} F(t)e^{-itx}dt = \int_0^{\infty} t\, e^{-t(1+ix)}dt = \frac{1}{(1+ix)^2} = 2\pi\ f(x),$$

say, so that

$$\hat{F}(x) = 2\pi\ f(-x).$$

Since $F, \hat{F} \in L_1(-\infty,\infty)$, and F is continuous, we have (by Theorem 11'),

$$F(t) = \frac{1}{2\pi}\int_{-\infty}^{\infty} \hat{F}(x)e^{-itx}dx = \int_{-\infty}^{\infty} f(-x)e^{-itx}dx = \int_{-\infty}^{\infty} f(x)e^{itx}dx = \hat{f}(t),$$

and the function

$$f(x) = \frac{1}{2\pi}\,\frac{1}{(1+ix)^2}$$

satisfies (11.2).

(11.3) Given an interval $(-\infty,a]$, or $[a,\infty)$, where a is a real number, there exists a function $f \in L_1(-\infty,\infty)$, such that its Fourier transform vanishes on the given interval, and does *not* vanish outside that interval.

For if we consider $\hat{h}(\alpha)$ defined by $\hat{h}(\alpha) = \hat{g}(\alpha-a)$, where $\hat{g}$ is defined by (11.2), then $\hat{h}$ is the Fourier transform of a function $h \in L_1$, and $\hat{h}$ vanishes on the interval $(-\infty,a]$ but not outside.

Similarly the Fourier transform of h(-t) is $\hat{h}(-\alpha) = \hat{g}(a-\alpha) = 0$, for $\alpha \geq a$, and $\hat{h}(-\alpha) \neq 0$, for $\alpha < a$.

Theorem 19. *Let $f \in L_1(-\infty,\infty)$, $\hat{f}(0) = 0$, and $\varepsilon > 0$. Then there exists a function $h \in L_1(-\infty,\infty)$, such that (i) $\|h\|_1 < \varepsilon$, (ii) $\hat{h} = \hat{f}$ in a neighbourhood of the origin, and (iii) $\hat{f}(\alpha) = 0$ implies that $\hat{h}(\alpha) = 0$.*

Proof. By (11.1) there exists a function $\lambda \in L_1(-\infty,\infty)$, such that $\hat{\lambda}(x) = 1$, for $|x| \leq 1$. If we set, for any fixed $R > 0$, $\lambda_R(x) = R\lambda(Rx)$, then

$$\hat{\lambda}_R(\alpha) = \hat{\lambda}(\tfrac{\alpha}{R}) = 1, \quad \text{for } |\alpha| \leq R,$$

since

$$R \int_{-\infty}^{\infty} \lambda(Rx)e^{i\alpha x}dx = \int_{-\infty}^{\infty} \lambda(y)e^{i\alpha y/R}dy .$$

Since $\hat{f}(0) = 0$ by hypothesis, we have $\int_{-\infty}^{\infty} f(x)dx = 0$, and

$$(\lambda_R*f)(x) = \int_{-\infty}^{\infty} \lambda_R(x-y)f(y)dy - \lambda_R(x)\int_{-\infty}^{\infty} f(y)dy$$

$$= \int_{-\infty}^{\infty} f(y)\left\{\lambda_R(x-y) - \lambda_R(x)\right\}dy.$$

Hence

$$\|\lambda_R*f\|_1 \leq \int_{-\infty}^{\infty} dx \int_{-\infty}^{\infty} |f(y)|\,|\lambda_R(x-y) - \lambda_R(x)|dy$$

$$= \int_{-\infty}^{\infty} |f(y)|dy \int_{-\infty}^{\infty} |\lambda_R(x-y) - \lambda_R(x)|dx \text{ (by Fubini's theorem)}$$

$$= \int_{-\infty}^{\infty} |f(y)|dy \int_{-\infty}^{\infty} R|\lambda(Rx - Ry) - \lambda(Rx)|dx$$

$$= \int_{-\infty}^{\infty} |f(y)|dy \int_{-\infty}^{\infty} |\lambda(x-Ry) - \lambda(x)|dx.$$

Now

$$\int_{-\infty}^{\infty} |\lambda(x-Ry) - \lambda(x)|dx \leq 2\|\lambda\|_1 ,$$

for every fixed y, and the (L_1-modulus of continuity of λ) integral tends to zero as $R \to 0$ for any fixed y (see (1.17)). Now choose M so large that

$$\int_{-\infty}^{-M} |f(y)| \; 2\|\lambda\|_1 \, dy < \varepsilon_1, \quad \int_{M}^{\infty} |f(y)| \; 2\|\lambda\|_1 \, dy < \varepsilon_2,$$

for any two strictly positive numbers $\varepsilon_1, \varepsilon_2$, given in advance. If $|y| \le M$, then $|yR| \le MR$, and $yR \to 0$ as $R \to 0$. Given $\varepsilon_3 > 0$ and $M > 0$, we can therefore choose R sufficiently small to ensure that

$$\int_{-M}^{M} |f(y)| \, dy \int_{-\infty}^{\infty} |\lambda(x-Ry) - \lambda(x)| dx < \varepsilon_3.$$

Thus we have $\|\lambda_R * f\|_1 < \varepsilon$, for any arbitrary $\varepsilon > 0$ given in advance, by proper choice of $R > 0$. With such an R we define $h = \lambda_R * f$, so that $\hat{h} = \hat{\lambda}_R \cdot \hat{f}$. By the definition of λ, we have $\hat{\lambda}_R(\alpha) = 1$ for $|\alpha| \le R$, hence $\hat{h}(x) = \hat{f}(x)$ for $|x| \le R$. Obviously $\hat{h}(\alpha) = 0$ if $\hat{f}(\alpha) = 0$.

Theorem 20. If $R > 0$, *and* $\varphi(z)$ *is holomorphic for* $|z| < R$, *with* $\varphi(0) = 0$, *and* $h \in L_1(-\infty,\infty)$, *with* $\|h\|_1 < R$, *then there exists a function* $g \in L_1(-\infty,\infty)$, *such that* $\varphi(\hat{h}) = \hat{g}$.

Proof. Let $f \in L_1(-\infty,\infty)$. Then the Fourier transform of the convolution of f with itself n times is the nth power of $\hat{f}$ (Theorem 4), and since the functional $f \to \hat{f}$ is linear, it follows that the function

$$P(z) = \sum_{k=1}^{n} a_k \, z^k, \qquad (z, \; a_k \text{ complex})$$

carries Fourier transforms into Fourier transforms.

By hypothesis we have $\varphi(z) = \sum_{n=1}^{\infty} a_n \, z^n$, $|z| < R$, the series converging absolutely ($a_0 = 0$ since $\varphi(0) = 0$). Since $\|h\|_1 < R$ by assumption, we have $|\hat{h}(x)| \le \|h\|_1 < R$, for all x, so that

$$\varphi(\hat{h}(x)) = \sum_{n=1}^{\infty} a_n \, (\hat{h}(x))^n, \quad -\infty < x < \infty.$$

Set $h_1 = h$, $h_k = h_{k-1} * h$, for $k \ge 2$. Then obviously $\|h_k\| \le \|h\|_1^k$, for $k \ge 2$, while $\hat{h}_k(x) = (\hat{h}(x))^k$, by (2.3) and (2.2).

If we choose integers m,n, such that $n \ge m \ge 1$, then

$$\left\| \sum_{k=m}^{n} a_k h_k \right\|_1 \le \sum_{k=m}^{n} |a_k| \, \|h_k\|_1 \le \sum_{k=m}^{n} |a_k| \, \|h\|_1^k .$$

Since $\|h\|_1 < R$, the series $\sum_{k=1}^{\infty} |a_k| \, \|h\|_1^k$ converges, so that $\sum_{k=m}^{n} |a_k| \, \|h\|_1^k \to 0$, as $m,n \to \infty$. Hence

$$\left\| \sum_{k=m}^{n} a_k h_k \right\|_1 \to 0, \text{ as } m,n \to \infty.$$

Since the function space $L_1(-\infty,\infty)$ is *complete* (if $\|f_m - f_n\|_1 \to 0$ as $m,n \to \infty$, then there exists $f \in L_1(-\infty,\infty)$, such that $\|f - f_n\|_1 \to 0$ as $n \to \infty$), there exists a function $g \in L_1(-\infty,\infty)$, such that

$$\left\| \sum_{k=1}^{n} a_k h_k - g \right\|_1 \to 0, \quad \text{as } n \to \infty.$$

Hence

$$\sum_{k=1}^{n} a_k \hat{h}_k(x) \to \hat{g}(x), \text{ as } n \to \infty,$$

uniformly in $-\infty < x < \infty$ (see (1.12)). Thus we have

$$\hat{g}(x) = \sum_{k=1}^{\infty} a_k \hat{h}_k(x) = \sum_{k=1}^{\infty} a_k (\hat{h}(x))^k = \varphi(\hat{h}(x)), \text{ for } -\infty < x < \infty.$$

(11.4) <u>*Corollary*</u>. *If φ is an entire function, with $\varphi(0) = 0$, then φ carries Fourier transforms (of functions in $L_1(-\infty,\infty)$) into Fourier transforms (of functions in $L_1(-\infty,\infty)$).*

<u>Remarks</u>

1. The condition $\varphi(0) = 0$ in Theorem 20 is necessary. For if $Q(z) = a_0 + P(z) = a_0 + \sum_{k=1}^{n} a_k z^k$, with $a_0 \neq 0$, then by Theorem 20, and the Riemann-Lebesgue theorem (Th.1), we have

$$\lim_{|x| \to \infty} P(\hat{f})(x) = 0,$$

while $Q(\hat{f}) = a_0 + P(\hat{f})$, and

$$\lim_{|x| \to \infty} Q(\hat{f})(x) = a_0 + \lim_{|x| \to \infty} P(\hat{f})(x) = a_0 \neq 0,$$

so that $Q(\hat{f})$ cannot be the Fourier transform of a function in $L_1(-\infty,\infty)$.

2. Given a finite interval $[a,b]$, there exists a function $\delta \in L_1(-\infty,\infty)$, such that $\hat{\delta}(\alpha) = 1$ for $a \le \alpha \le b$ (see (11.1)). Thus, if $f \in L_1(-\infty,\infty)$, then

$$Q(\hat{f})(x) = a_0 \hat{\delta}(x) + P(\hat{f})(x), \text{ for } a \le x \le b.$$

Hence $Q(\hat{f})$ coincides on a bounded, closed interval $[a,b]$ with the Fourier transform $a_0\hat{\delta} + P(\hat{f})$ of a function in $L_1(-\infty,\infty)$. More generally we have

Theorem 21. *Let D be a domain (that is, an open, connected set) in the complex plane, and φ a function holomorphic in D. Let $f \in L_1(-\infty,\infty)$, and $\hat{f}(x) \in D$ for $-\infty < a \le x \le b < \infty$. Then there exists a function $g \in L_1(-\infty,\infty)$, such that $\varphi(\hat{f}(x)) = \hat{g}(x)$, for $a \le x \le b$.*

For the proof we shall use two lemmas.

(11.4) Lemma. *If $f \in L_1(-\infty,\infty)$, $\hat{f}(0) = 0$, and φ is holomorphic at the origin, and $\varphi(0) = 0$, then there exists a function $g \in L_1(-\infty,\infty)$, such that $\varphi(\hat{f}) = \hat{g}$ in a neighbourhood of the origin.*

Proof. There exists a number $\varepsilon > 0$, such that φ is holomorphic in $|z| < \varepsilon$. Hence, by Theorem 19, there exists a function $h \in L_1(-\infty,\infty)$, such that $\|h\|_1 < \varepsilon$, and

$$\hat{f}(x) = \hat{h}(x), \quad x \in N_0,$$

where N_0 is a neighbourhood of the origin. By Theorem 20 there exists a function $g \in L_1(-\infty,\infty)$, such that $\varphi(\hat{h}) = \hat{g}$ in $(-\infty,\infty)$; in particular, $\varphi(\hat{f}(x)) = \hat{g}(x)$, for $x \in N_0$.

(11.5) Lemma. *If $f \in L_1(-\infty,\infty)$, $\hat{f}(\alpha) = \beta$, and φ is holomorphic at β, then there exixts a function $g \in L_1(-\infty,\infty)$, such that*

$$\varphi(\hat{f}(x)) = \hat{g}(x), \quad x \in N_\alpha,$$

where N_α is a neighbourhood of α.

Proof. If the lemma holds for $\alpha = 0$, then it holds for arbitrary real α. For suppose that $\alpha \neq 0$, and set $f_1(t) = e^{i\alpha t}f(t)$, $-\infty < t < \infty$. Then $\hat{f}_1(x) = \hat{f}(x+\alpha)$, therefore $\hat{f}_1(0) = \hat{f}(\alpha) = \beta$. If we assume the lemma true for $\alpha = 0$, then there exists a function $g_1 \in L_1(-\infty,\infty)$, such that $\varphi(\hat{f}_1(x)) = \hat{g}_1(x)$, for all x in a neighbourhood of the origin. Therefore $\varphi(\hat{f}_1(x-\alpha)) = \hat{g}_1(x-\alpha)$, for all x in a neighbourhood of α, say N_α. If we define g by the condition $\hat{g}(\alpha) = \hat{g}_1(x-\alpha)$, then we have

$$\varphi(\hat{f}(x)) = \hat{g}(x), \ x \in N_\alpha,$$

since $\hat{f}_1(x-\alpha) = \hat{f}(x)$.

If the lemma holds for $\beta = 0$, then it holds also for $\beta \neq 0$. For suppose that $\beta \neq 0$, and set $\hat{f}_1(x) = \hat{f}(x) - \beta\hat{\delta}(x)$, where $\hat{\delta}(x) = 1$ for $|x| \leq 1$ (as in (11.1)), and let $\psi(z) = \varphi(z+\beta)$. Then $\hat{f}_1(0) = \hat{f}(0) - \beta\hat{\delta}(0) = \beta-\beta = 0$, and $\psi(z)$ is holomorphic at $z = 0$. By the assumption that the lemma holds for $\beta = 0$, there exists a function $g \in L_1(-\infty,\infty)$, such that

$$\psi(\hat{f}_1(x)) = \hat{g}(x), \ x \in N_0,$$

where N_0 is a neighbourhood of the origin. We may assume that $N_0 \subset [-1,1]$. Then we have, for $x \in N_0$,

$$\varphi(\hat{f}(x)) = \psi(\hat{f}(x) - \beta) = \psi(\hat{f}(x) - \beta\hat{\delta}(x)) = \psi(\hat{f}_1(x)) = \hat{g}(x),$$

which proves the lemma in the case $\beta \neq 0$.

If the lemma holds in the case $\varphi(0) = 0$, then it holds also in the case $\varphi(0) \neq 0$. For suppose that $\varphi(0) \neq 0$, and set $\psi(z) = \varphi(z) - \varphi(0)$, so that $\psi(0) = 0$. On the assumption that the lemma holds in the case $\varphi(0) = 0$, we can conclude that there exists a function $g_1 \in L_1(-\infty,\infty)$, such that $\psi(\hat{f}(x)) = \hat{g}_1(x)$, $x \in N_0$, where N_0 is a neighbourhood of the origin. If we define g by the condition $\hat{g} = \hat{g}_1 + \varphi(0)\hat{\delta}$, (g exists since $\hat{g}_1 + \varphi(0)\hat{\delta}$ is the Fourier transform of a function in $L_1(-\infty,\infty)$), then we have, for $x \in N_0$,

$$\varphi(\hat{f}(x)) = \psi(\hat{f}(x)) + \varphi(0) = \psi(\hat{f}(x)) + \varphi(0)\hat{\delta}(x)$$

$$= \hat{g}_1(x) + \varphi(0)\hat{\delta}(x) = \hat{g}(x).$$

The proof of Lemma (11.5) is thus reduced to that of Lemma (11.4), which is already proved.

Proof of Theorem 21. By Lemma (11.5), every point $x \in [a,b]$ is covered by an open interval on which $\varphi(\hat{f})$ coincides with the Fourier transform of a function in $L_1(-\infty,\infty)$. By the theorem of Heine-Borel, a finite number of such intervals cover $[a,b]$. We may suppose that none of those intervals is wholly contained in another. Let (α_1,β_1) and (α_2,β_2) be two of those intervals. We may suppose that

$$\alpha_1 < \alpha_2 < \beta_1 < \beta_2.$$

We choose $g_1, g_2 \in L_1(-\infty,\infty)$, such that

$$\varphi(\hat{f}(x)) = \hat{g}_1(x), \quad \alpha_1 < x < \beta_1,$$

and

$$\varphi(\hat{f}(x)) = \hat{g}_2(x), \quad \alpha_2 < x < \beta_2.$$

It follows that $\hat{g}_1(x) = \hat{g}_2(x)$ for $\alpha_2 < x < \beta_1$; in fact, for $\alpha_2 \leq x \leq \beta_1$, since the Fourier transform of a function in $L_1(-\infty,\infty)$ is continuous.

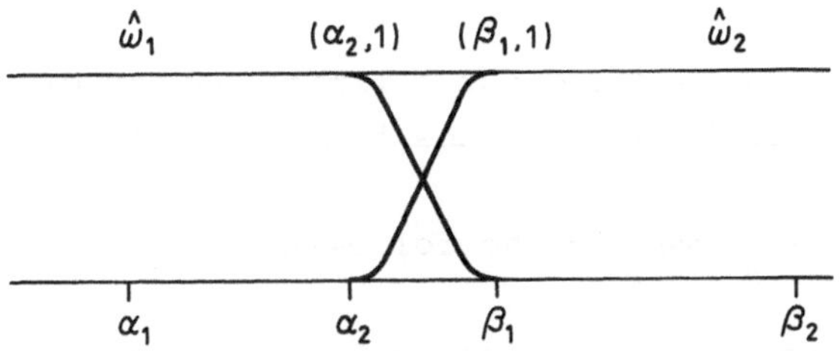

As in (3.12), and (11.1), there exist infinitely differentiable functions ω_1, ω_2, such that

$$\hat{\omega}_1(x) = \begin{cases} 0, & x \geq \beta_1 \\ 1, & \text{for } \alpha_1 \leq x \leq \alpha_2 \\ \frac{1}{c}\int_x^{\beta_1} e^{\frac{1}{(y-\beta_1)(y-\alpha_2)}}\,dy, \quad c = \int_{\alpha_2}^{\beta_1} e^{\frac{1}{(y-\beta_1)(y-\alpha_2)}}\,dy, & \text{for } \alpha_2 \leq x \leq \beta_1 \end{cases}$$

$$\hat{\omega}_2(x) = \begin{cases} 0, & x \le \alpha_2, \\ 1, & \text{for } \beta_1 \le x \le \beta_2 \\ \frac{1}{c}\int_{\alpha_2}^{x} e^{\frac{1}{(y-\alpha_2)(y-\beta_1)}} dy, & \text{for } \alpha_2 \le x \le \beta_1. \end{cases}$$

For $\alpha_2 \le x \le \beta_1$, we have therefore $\hat{\omega}_1(x) + \hat{\omega}_2(x) = 1$.

If we *define* $\hat{\psi} = \hat{\omega}_1\hat{g}_1 + \hat{\omega}_2\hat{g}_2$, then we have, for $x \in (\alpha_1,\alpha_2)$

$$\hat{\psi}(x) = \hat{g}_1(x) = \varphi(\hat{f}(x)),$$

and if $x \in (\beta_1,\beta_2)$, then

$$\hat{\psi}(x) = \hat{g}_2(x) = \varphi(\hat{f}(x)).$$

If $x \in [\alpha_2,\beta_1]$, then $\hat{g}_1(x) = \hat{g}_2(x)$, so that

$$\hat{\psi}(x) = [\hat{\omega}_1(x)+\hat{\omega}_2(x)]\,\hat{g}_1(x) = \hat{g}_1(x) = \varphi(\hat{f}(x))$$

(by choice of $\hat{g}_1$). Hence

$$\hat{\psi}(x) = \varphi(\hat{f}(x)), \text{ for } \alpha_1 < x < \beta_2.$$

This argument can be repeated for a finite number of such intervals, which together cover [a,b], and the theorem follows.

By choosing $\varphi(z) = \frac{1}{z}$, we get the following

(11.6) <u>*Corollary*</u> *(Wiener). Let* $f \in L_1(-\infty,\infty)$*, and let* $\hat{f}(\alpha) \neq 0$*, for* $x \in [a,b]$*. Then there exists a function* $g \in L_1(-\infty,\infty)$*, such that*

$$\frac{1}{\hat{f}(\alpha)} = \hat{g}(\alpha), \text{ for } \alpha \in [a,b].$$

§12. The closure of translations

Let $f \in L_1(-\infty,\infty)$, and let S_f denote the set of finite linear combinations of translations of f, that is to say

$$S_f = \left\{ \sum_{k=1}^{m} a_k f(\cdot + c_k) \right\},$$

where c_k is real, a_k complex, and m an integer, $m \geq 1$. Clearly we have $S_f \subset L_1(-\infty,\infty)$. Let $\overline{S}_f$ denote the closure, in the L_1-norm, of the set S_f. Then $g \in \overline{S}_f$ if and only if there exists a sequence $(g_n), g_n \in S_f$, such that $\|g-g_n\|_1 \to 0$ as $n \to \infty$. Obviously $S_f \subset \overline{S}_f$, and since L_1 is *complete*, we have $\overline{S}_f \subset L_1$. Clearly $\overline{\overline{S}}_f = \overline{S}_f$, since $\overline{S}_f$ is closed.

The set $\overline{S}_f$ is linear, in the sense that if $h_1, h_2 \in \overline{S}_f$, then $a_1h_1 + a_2h_2 \in \overline{S}_f$, for any two complex numbers a_1, a_2. The set $\overline{S}_f$ is translation invariant, in the sense that if $h(\cdot) \in \overline{S}_f$, then $h(\cdot + c) \in \overline{S}_f$, for any real c. Finally, if $h \in \overline{S}_f$, and we consider the closure (of the set of finite linear combinations) of translations of h, namely $\overline{S}_h$, then $\overline{S}_h \subset \overline{S}_f$, so that we do not obtain any functions not already contained in $\overline{S}_f$. For let $F \in \overline{S}_h$. Then there exists a finite linear combination of translations of h, say $\sum_{k=1}^{n} a_k h(x + c_k)$, such that for any given $\varepsilon > 0$,

$$\|F - \sum_{k=1}^{n} a_k h(\cdot + c_k)\|_1 < \varepsilon.$$

If $h \in \overline{S}_f$, then $a_k h(\cdot + c_k) \in \overline{S}_f$, hence $\sum_{k=1}^{n} a_k h(\cdot + c_k) \in \overline{S}_f$, so that $\sum_{k=1}^{n} a_k h(\cdot + c_k)$ can be approximated in the L_1-norm by translations of f, and hence also F, so that $F \in \overline{S}_f$.

<u>*Theorem 2.2*</u> *(Wiener). If* $f \in L_1(-\infty,\infty)$, *and* $\overline{S}_f = L_1(-\infty,\infty)$, *then* $\hat{f}(x) \neq 0$, $-\infty < x < \infty$.

<u>Proof</u>. Suppose the theorem false. Then there exists a real number a, say, such that $\hat{f}(a) = 0$. By (11.1) there exists a function $\delta \in L_1(-\infty,\infty)$, such that $\hat{\delta}(a) = 1$. We will show that if $\delta \in \overline{S}_f = L_1$, then $\hat{\delta}(a) = 0$, which is a contradiction.

For if $f \in L_1(-\infty,\infty)$, and $\hat{f}(a) = 0$, then $\hat{g}(a) = 0$ for *every* $g \in \overline{S}_f$. To prove this, take any $h \in S_f$. Then $h(x) = \sum a_k f(x + c_k)$, and $\hat{h}(x) = \sum a_k \hat{f}(x) e^{-ic_k x}$. If $\hat{f}(a) = 0$, then $\hat{h}(a) = 0$. If now $g \in \overline{S}_f$, then there exists a sequence $(g_n), g_n \in S_f$, such that $\|g-g_n\|_1 \to 0$, as $n \to \infty$. Taking $h = g_n$, we see that $\hat{g}_n(a) = 0$ for *every* n. Therefore $\hat{g}(a) = \lim_{n\to\infty} \hat{g}_n(a) = 0$. (Note that if $\|g_n-g\|_1 \to 0$, then, by (1.12),

$\hat{g}_n(\alpha) \to \hat{g}(\alpha)$, *uniformly* in α).

If $\overline{S}_f = L_1$, then $\delta \in \overline{S}_f$, and by what we have just proved, $\hat{\delta}(a) = 0$, whereas $\hat{\delta}(a) = 1$, which is a contradiction.

Theorem 23 (Wiener). If $f \in L_1(-\infty,\infty)$, *and if* $\hat{f}(\alpha) \neq 0$, $-\infty < \alpha < \infty$, *then* $\overline{S}_f = L_1(-\infty,\infty)$.

In the arrangement of the proof, we follow Bochner, and prove three lemmas first.

(12.1) *Lemma. If* $f \in L_1(-\infty,\infty)$, $g \in \overline{S}_f$, $h \in L_1(-\infty,\infty)$, *then* $g*h \in \overline{S}_f$.

Proof. We may assume that neither g nor h is almost everywhere zero. (Note that $g \in L_1(-\infty,\infty)$, since $\overline{S}_f \subset L_1$).

Let $H = g*h$, so that

$$H(x) = \int_{-\infty}^{\infty} g(x-t)h(t)\,dt,$$

where the integral exists for almost all x, and belongs to $L_1(-\infty,\infty)$ (see Lemma (2.1)).

Given $\varepsilon > 0$, choose N so large that

$$\int_{|x|\geq N} |h(x)|\,dx < \frac{\varepsilon}{2\|g\|_1}. \tag{12.2}$$

Let

$$H_N(x) = \int_{-N}^{N} g(x-t)h(t)\,dt,$$

so that

$$H(x) - H_N(x) = \int_{|t|\geq N} g(x-t)h(t)\,dt,$$

and

$$\|H-H_N\|_1 \leq \int_{-\infty}^{\infty} dx \int_{|t|\geq N} |g(x-t)|\ |h(t)|\,dt$$

$$= \int_{|t|\geq N} |h(t)|dt \int_{-\infty}^{\infty} |g(x-t)|dx$$

$$(12.3) \qquad = \int_{|t|\geq N} |h(t)|dt \; ||g||_1 < \frac{\varepsilon}{2}$$

by (12.2).

Given $\varepsilon > 0$, there exists $\delta > 0$, such that

$$(12.4) \qquad \int_{-\infty}^{\infty} |g(x-y) - g(x)|dx \leq \frac{\varepsilon}{2||h||_1}, \text{ for } |y| \leq \delta$$

(see (1.17)). Choose a finite sequence $t_1,\dots,t_n$ such that

$$-N = t_0 < t_1 < \dots < t_n = N$$

and such that $t_k - t_{k-1} \leq \delta$, for $1 \leq k \leq n$, where n remains at our disposal. Then

$$H_N(x) = \sum_{k=1}^{n} \int_{t_{k-1}}^{t_k} g(x-t)h(t)dt.$$

If we now define

$$h_N(x) = \sum_{k=1}^{n} g(x-t_k) \int_{t_{k-1}}^{t_k} h(t)dt,$$

then clearly $h_N \in S_g$, and we have

$$H_N(x) - h_N(x) = \sum_{k=1}^{n} \int_{t_{k-1}}^{t_k} [g(x-t) - g(x-t_k)]h(t)dt ,$$

hence

$$||H_N - h_N||_1 \leq \sum_{k=1}^{n} \int_{-\infty}^{\infty} dx \int_{t_{k-1}}^{t_k} |g(x-t)-g(x-t_k)| \; |h(t)|dt$$

$$= \sum_{k=1}^{n} \int_{t_{k-1}}^{t_k} |h(t)|dt \int_{-\infty}^{\infty} |g(x-t)-g(x-t_k)|dx.$$

If $t_{k-1} \leq t \leq t_k$, then $|t-t_k| \leq \delta$. And for $|t-t_k| \leq \delta$, we have, by (12.4),

$$\int_{-\infty}^{\infty} |g(x-t) - g(x-t_k)| \leq \frac{\varepsilon}{2||h||_1} ,$$

so that

(12.5) $$\|H_N-h_N\|_1 \le \frac{\varepsilon}{2\|h\|_1} \sum_{k=1}^{n} \int_{t_{k-1}}^{t_k} |h(t)|\,dt = \frac{\varepsilon}{2\|h\|_1} \int_{-N}^{N} |h(t)|\,dt \le \frac{\varepsilon}{2},$$

(since $h \not\equiv 0$), and

$$\|H-h_N\|_1 \le \|H-H_N\|_1 + \|H_N-h_N\|_1 < \varepsilon,$$

by (12.3) and (12.5). Since $h_N \in S_g$, it follows that $H \in \overline{S}_g$. By hypothesis $g \in \overline{S}_f$, hence $\overline{S}_g \subset \overline{S}_f$ (as noted before the enunciation of the theorem), therefore $H \in \overline{S}_f$.

(12.6) Lemma. *Let* $f \in L_1(-\infty,\infty)$, *and let*

$$H_R(t) = R\left(\frac{\sin(Rt/2)}{(Rt/2)}\right)^2, \quad R > 0, \ -\infty < t < \infty.$$

If $H_R \in \overline{S}_f$ *for* $R = 1,2,\ldots,$ *then* $\overline{S}_f = L_1(-\infty,\infty)$.

Proof. Take any function $h \in L_1(-\infty,\infty)$. *If* $H_R \in \overline{S}_f$ for $R = 1,2,\ldots,$ then by the preceding Lemma (12.1), we have $H_R*h \in \overline{S}_f$ for $R = 1,2,\ldots$. By Theorem 16, (9.3), we know that

$$\left\|\frac{1}{2\pi}(H_R*h) - h\right\|_1 \to 0, \text{ as } R \to \infty.$$

Hence $h \in \overline{\overline{S}}_f = \overline{S}_f$, for every $h \in L_1(-\infty,\infty)$, which implies that $L_1(-\infty,\infty) \subset \overline{S}_f$. But $\overline{S}_f \subset L_1(-\infty,\infty)$, and the lemma follows.

(12.7) Lemma. *If* $f \in L_1(-\infty,\infty)$, *and* $\hat{f}(x) \neq 0$, $-\infty < x < \infty$, *then*

$$H_R \in \overline{S}_f, \quad \textit{for } R = 1,2,\ldots\ .$$

Proof. Let $\infty > R > 0$, and $\hat{f}(x) \neq 0$ for $-R \le x \le R$. By Corollary (11.6) of Theorem 21, there exists a function $g \in L_1(-\infty,\infty)$, such that $\frac{1}{\hat{f}(x)} = \hat{g}(x)$, for $-R \le x \le R$. If $\hat{H}_R(x) = 2\pi\, K_R(x)$, (see (8.1) as well as the Remarks following Theorem 15, §9), then $K_R(x) = 0$ for $|x| \ge R$, and

$$\frac{2\pi\, K_R(x)}{\hat{f}(x)} = 2\pi\, K_R(x)\, \hat{g}(x),$$

hence

$$2\pi\, K_R(x) = 2\pi\, \hat{f}(x)\, K_R(x)\, \hat{g}(x).$$

But $2\pi\, \hat{f}\, K_R\, \hat{g}$ is the Fourier transform of $f*H_R*g$ (Theorem 2, §2). Hence by the uniqueness theorem (Theorem 7, §6), we have

$$H_R = f*H_R*g = f*(H_R*g),$$

where $H_R*g \in L_1(-\infty,\infty)$, and $f \in \overline{S}_f$. By Lemma 12.1, it follows that $H_R \in \overline{S}_f$.

<u>Proof of Theorem 23.</u> If $\hat{f}(x) \neq 0$, $-\infty < x < \infty$, then by the immediately preceding Lemma (12.7), $H_R \in \overline{S}_f$, for $R = 1,2,\ldots,n,\ldots$, hence, by Lemma (12.6), $\overline{S}_f = L_1(-\infty,\infty)$.

§13. A general tauberian theorem

A tauberian theorem is the corrected converse of an abelian theorem. The word "abelian", in this context, originates from Abel's theorem on power series, which states that if (a_n) is an infinite sequence of real numbers, and $\sum_{n=0}^{\infty} a_n$ converges, and has the sum s, then the power series $\sum_{n=0}^{\infty} a_n x^n$ converges uniformly for $0 \leq x \leq 1$, and

$$\lim_{x\uparrow 1} \sum_{n=0}^{\infty} a_n x^n = s.$$

The direct converse of this theorem is false. If we take $a_n = (-1)^n$, $n = 1,2,\ldots$, then $\sum_{n=0}^{\infty} a_n x^n = \frac{1}{1+x}$, $0 \leq x < 1$ and $\lim_{x\uparrow 1} \sum_{n=0}^{\infty} (-1)^n x^n = \frac{1}{2}$, but $\sum_{n=0}^{\infty} a_n$ is *not* convergent. Tauber proved (1897) that the converse is correct under the additional condition, known ever since as a "tauberian condition", that $n\, a_n \to 0$, $n \to \infty$. Tauber's theorem was later sharpened by J.E. Littlewood, who showed (1910) that the condition $n\, a_n = O(1)$, as $n \to \infty$, was sufficient to prove the converse of Abel's theorem, and thereby provided the impetus for the remarkable work of Hardy and Littlewood on a variety of special problems. Adopting a totally different point of view, Wiener showed (1930-32) that "most" tauberian theorems, like the converse of Abel's theorem, follow as special cases of a "general tauberian theorem", which properly belongs to the theory of Fourier transforms on $L_1(-\infty,\infty)$, and of which the following is the simplest version.

Theorem 24 (Wiener). Let $h(x)$ *be a bounded (measurable) function defined for* $-\infty < x < \infty$*, and let* $K_1(x) \in L_1(-\infty < x < \infty)$*, with* $\hat{K}_1(\alpha) \neq 0$ *for* $-\infty < \alpha < \infty$*. If*

$$(13.1) \qquad \lim_{x\to\infty} \int_{-\infty}^{\infty} K_1(x-\xi)h(\xi)d\xi = A \int_{-\infty}^{\infty} K_1(\xi)d\xi, \text{ for some complex } A,$$

then we have, for every $K \in L_1(-\infty,\infty)$*,*

$$(13.2) \qquad \lim_{x\to\infty} \int_{-\infty}^{\infty} K(x-\xi)h(\xi)d\xi = A \int_{-\infty}^{\infty} K(\xi)d\xi.$$

The following is a kind of converse, which is easier to prove.

Theorem 25 (Wiener). Let $K_1 \in L_1(-\infty,\infty)$*, and let its Fourier transform* $\hat{K}_1$ *have a real zero. Then there exist a bounded (measurable) function* h *on* $(-\infty,\infty)$*, and a function* $K \in L_1(-\infty,\infty)$*, such that (13.1) is true, but (13.2) is false.*

Proof. If $\hat{K}_1(c) = 0$ for a certain real c, choose $K \in L_1(-\infty,\infty)$, such that $\hat{K}(c) \neq 0$, for example $K(x) = e^{-(1/2)x^2}$ (Example 6, §1), and let $h(\xi) = e^{-ic\xi}$. Then we have

$$\int_{-\infty}^{\infty} K_1(x-\xi)e^{-ic\xi}d\xi = \int_{-\infty}^{\infty} K_1(t)e^{ic(t-x)}dt = e^{-icx}\,\hat{K}_1(c) = 0,$$

for every real x. But

$$\int_{-\infty}^{\infty} K(x-\xi)e^{-ic\xi}d\xi = e^{-icx}\,\hat{K}(c) \not\to \text{ a limit, as } x\to\infty.$$

Proof of Theorem 24. If $K(x)$ is of the form $K(x) = \sum_{k=1}^{n} A_k K_1(x+\lambda_k)$, where A_k is complex, and λ_k real, then the theorem is trivially true. By Theorem 23 on the L_1-closure of translations, given any $K \in L_1(-\infty,\infty)$, there exists a function K_3 of the form

$$K_3(x) = \sum_{k=1}^{n} A_k K_1(x+\lambda_k),$$

such that

$$(13.3) \qquad \int_{-\infty}^{\infty} |K(x) - K_3(x)|dx < \varepsilon,$$

where ε is any strictly positive number given in advance. If $|h(x)| < B < \infty$, then we have

$$(13.4)\qquad \left|\int_{-\infty}^{\infty} K(x-\xi)h(\xi)d\xi - \int_{-\infty}^{\infty} K_3(x-\xi)h(\xi)d\xi\right| < B\cdot\varepsilon.$$

Since (13.2) holds with $K = K_3$, we have

$$(13.5)\qquad \left|\int_{-\infty}^{\infty} K_3(x-\xi)h(\xi)d\xi - A\int_{-\infty}^{\infty} K_3(\xi)d\xi\right| < \varepsilon, \quad x \geq x_0 > 0,$$

while (13.3) gives

$$(13.6)\qquad \left|A\int_{-\infty}^{\infty} K_3(\xi)d\xi - A\int_{-\infty}^{\infty} K(\xi)d\xi\right| < |A|\varepsilon.$$

Combining (13.4), (13.5), and (13.6), we obtain

$$\limsup_{x\to\infty} \left|\int_{-\infty}^{\infty} K(x-\xi)h(\xi)d\xi - A\int_{-\infty}^{\infty} K(\xi)d\xi\right| \leq B\varepsilon + \varepsilon + |A|\varepsilon,$$

and hence the theorem.

<u>*Theorem 26*</u> *(J.E. Littlewood). Let* $f(x) = \sum_{n=0}^{\infty} a_n x^n$, *for* $|x| < 1$, *where* a_n *is complex, and let*

$$\lim_{x\uparrow 1} f(x) = s, \qquad (s \textit{ finite}).$$

If $a_n = O(1/n)$, *then* $\sum_{n=0}^{\infty} a_n = s$.

<u>Proof</u>. Let

$$S(x) = \sum_{n=0}^{[x]} a_n, \quad x > 0,$$

where [x] denotes the greatest integer not exceeding x. Then we have

$$S(x) - f(e^{-1/x}) = \sum_{n=0}^{[x]} a_n - \sum_{m=0}^{\infty} a_m e^{-m/x}$$

$$= \sum_{n=1}^{[x]} a_n(1-e^{-n/x}) + \sum_{m=[x]+1}^{\infty} a_m e^{-m/x},$$

and since $n|a_n| < M < \infty$, for all $n \geq 1$, and $[x] \leq x$, $[x]+1 > x$, we have

$$(13.7)\qquad |S(x) - f(e^{-1/x})| \leq \sum_{n=1}^{[x]} \frac{M}{n}\cdot\frac{n}{x} + O(1) = O(1).$$

By hypothesis, $f(e^{-1/x})$ is *bounded* as $x\to\infty$, hence (13.7) implies that

$$(13.8)\qquad S(x) = O(1), \quad \text{as } x\to\infty.$$

Consider the function $f(x) = \sum_{n=0}^{\infty} a_n x^n$, with $x = e^{-(e^{-\xi})}$:

$$f\left(e^{-e^{-\xi}}\right) = \sum_{n=0}^{\infty} a_n e^{-ne^{-\xi}} = \int_0^{\infty} e^{-te^{-\xi}} dS(t) = \int_0^{\infty} e^{-\xi} e^{-te^{-\xi}} S(t)dt$$

$$= \int_{-\infty}^{\infty} e^{-(\xi-\eta)} e^{-e^{-(\xi-\eta)}} S(e^{\eta})d\eta. \quad (t=e^{\eta})$$

We note that $\xi \to \infty$ as $x \uparrow 1$, and if we choose $K_1(\xi) = e^{-\xi} e^{-e^{-\xi}}$, then $K_1 \in L_1(-\infty,\infty)$, and

$$\hat{K}_1(\alpha) = \int_{-\infty}^{\infty} K_1(\xi) e^{i\alpha\xi}d\xi = \Gamma(1-i\alpha) \neq 0, \qquad \text{(Example 3, §1)}$$

so that $\int_{-\infty}^{\infty} K_1(\xi)d\xi = \Gamma(1) = 1$. Our hypotheses imply that

$$(13.9) \qquad \lim_{\xi\to\infty} \int_{-\infty}^{\infty} K_1(\xi-\eta) S(e^{\eta})d\eta = s = s \int_{-\infty}^{\infty} K_1(\eta)d\eta \quad .$$

Because of (13.8) this implies, by Theorem 25, that

$$(13.10) \qquad \lim_{\xi\to\infty} \int_{-\infty}^{\infty} K(\xi-\eta) S(e^{\eta})d\eta = s \int_{-\infty}^{\infty} K(\eta)d\eta \ ,$$

for *every* $K \in L_1(-\infty,\infty)$. If we choose for K the Littlewood function given by

$$K(\xi) = \begin{cases} 0, & \text{for } \xi \leq 0, \\ e^{-\xi}, & \text{for } 0 < \xi \leq L, \quad L > 0, \\ 0, & \text{for } \xi > L, \end{cases}$$

we see that

$$\int_{-\infty}^{\infty} K(\xi-\eta) S(e^{\eta})d\eta = \int_{\xi-L}^{\xi} e^{\eta-\xi}S(e^{\eta})d\eta = \frac{1}{x}\int_{xe^{-L}}^{x} S(y)dy,$$

where $x = e^{\xi}$, so that $x \to \infty$ as $\xi \to \infty$, and

$$\int_0^L K(\eta)d\eta = \int_0^L e^{-\eta}d\eta = (1-e^{-L}).$$

Hence (13.10) becomes

$$(13.11) \qquad \lim_{x\to\infty} \frac{1}{x}\int_{xe^{-L}}^{x} S(y)dy = s(1-e^{-L}).$$

Now for $xe^{-L} \leq y \leq x$, we have

$$|S(x) - S(y)| \le M \frac{x(1-e^{-L})+1}{xe^{-L}},$$

since

$$|S(x) - S(y)| = \left|\sum_{n=[y+1]}^{[x]} a_n\right| \le \frac{M}{y}[(x-y)+1] \le M \frac{x(1-e^{-L})+1}{xe^{-L}}.$$

On writing the identity

$$S(x)-s = \frac{1}{x(1-e^{-L})} \int_{xe^{-L}}^{x} \{S(x)-S(y)\}dy + \frac{1}{x(1-e^{-L})} \int_{xe^{-L}}^{x} S(y)dy - s,$$

and making use of (13.11), we obtain

$$\limsup_{x\to\infty} |S(x)-s| \le \lim_{x\to\infty} M \frac{x(1-e^{-L})+1}{xe^{-L}\cdot x(1-e^{-L})} x(1-e^{-L}) = M \frac{1-e^{-L}}{e^{-L}},$$

for every $L > 0$. On letting $L \downarrow 0$, we obtain the result: $S(x) \to s$, as $x \to \infty$.

§14. Two differential equations

To illustrate the application of Fourier transform methods in the study of differential equations, we shall consider two simple cases: the equation of heat conduction in an infinite rod, namely

(14.1) $$\frac{\partial u(x,t)}{\partial t} = \frac{\partial^2 u(x,t)}{\partial x^2},$$

under suitable conditions, and Laplace's equation

(14.2) $$\frac{\partial^2 v(x,t)}{\partial x^2} + \frac{\partial^2 v(x,t)}{\partial t^2} = 0,$$

which governs the distribution of temperature in an infinitely large plate, under suitable conditions. The first equation is connected with the *Gauss-Weierstrass integral* of $f \in L_1(-\infty,\infty)$, and the second with the *Cauchy-Poisson integral* of $f \in L_1(-\infty,\infty)$, studied in §9 (see Corollaries (7.16), (7.18), (9.5) and (9.6)). These are given respectively by

(14.3) $$U(f;x,t) \equiv U(x,t) = \int_{-\infty}^{\infty} f(\xi)W(x-\xi,t)d\xi, \quad f \in L_1(-\infty,\infty), \; t > 0,$$

$$W(x,t) = \frac{1}{2\sqrt{(\pi t)}} e^{-x^2/4t}, \quad -\infty < x < \infty,$$

and

(14.4) $$V(f;x,t) \equiv V(x,t) = \int_{-\infty}^{\infty} f(\xi)P(x-\xi,t)d\xi, \quad f \in L_1(-\infty,\infty), \; t > 0,$$

$$P(x,t) = \frac{1}{\pi} \frac{t}{t^2+x^2} .$$

We note that $W(x,t)$ satisfies (14.1), and is the so-called fundamental solution. Similarly $P(x,t)$ satisfies (14.2).

We shall identify those properties of $U(x,t)$ which will show that it is the unique solution (almost everywhere) of equation (14.1). We assume throughout that $f \in L_1(-\infty,\infty)$.

U_1. For each $t > 0$, $U(x,t) \in L_1(-\infty,\infty)$ as a function of x, and

$$\|U(\cdot,t)\|_1 \leq \|f\|_1 .$$

U_2. We have $$\lim_{t\to 0+} \|U(\cdot,t) - f(\cdot)\|_1 = 0,$$

as was proved in (9.5).

U_3. For $t > 0$, and $-\infty < x < \infty$, the partial derivatives $\frac{\partial U}{\partial x}, \frac{\partial^2 U}{\partial x^2}$ exist and belong to $L_1(-\infty,\infty)$ as functions of x.

This can be seen by differentiating under the integral sign and noting that the resulting integral is the convolution of two functions in $L_1(-\infty,\infty)$.

U_4. For $t > 0$, $\frac{\partial U}{\partial t} \in L_1(-\infty,\infty)$ as a function of x, and

$$\lim_{h\to 0} \left\| \frac{U(\cdot,t+h) - U(\cdot,t)}{h} - \frac{\partial U(\cdot,t)}{\partial t} \right\|_1 = 0 .$$

Since

$$U(x,t) = \int_{-\infty}^{\infty} f(\xi)W(x-\xi,t)d\xi, \quad f \in L_1(-\infty,\infty),$$

$$W(x,t) = \frac{1}{2\sqrt{(\pi t)}} e^{-x^2/4t} = W(-x,t), \quad t > 0,$$

we have for $h \neq 0$

(14.5)
$$\int_{-\infty}^{\infty} \left| \frac{U(x,t+h) - U(x,t)}{h} - \frac{\partial U(x,t)}{\partial t} \right| dx$$

$$\leq \int_{-\infty}^{\infty} dx \int_{-\infty}^{\infty} |f(x-\xi)| \left| \frac{W(\xi,t+h) - W(\xi,t)}{h} - \frac{\partial W(\xi,t)}{\partial t} \right| d\xi$$

$$\leq \|f\|_1 \int_{-\infty}^{\infty} \left| \frac{W(\xi,t+h) - W(\xi,t)}{h} - \frac{\partial W(\xi,t)}{\partial t} \right| d\xi .$$

But

$$\lim_{h\to 0} \frac{W(\xi,t+h) - W(\xi,t)}{h} = \frac{\partial W(\xi,t)}{\partial t}$$

for every ξ. This limit relation holds also in the L_1-norm, for

$$\frac{\partial W(\xi,t)}{\partial t} = \frac{1}{2\sqrt{\pi}} \left\{ -\frac{1}{2} t^{-3/2} e^{-\xi^2/4t} + \frac{\xi^2}{4} t^{-5/2} e^{-\xi^2/4t} \right\} ,$$

$t > 0$, so that

$$\left| \frac{\partial W(\xi,t)}{\partial t} \right| \leq \psi(\xi) \in L_1(-\infty < \xi < \infty),$$

for fixed $t > 0$, in the interval $[t-h,t+h]$, with h sufficiently small, and secondly

$$\frac{W(\xi,t+h) - W(\xi,t)}{h} = \frac{1}{h} \int_0^h W_1(\xi,t+y)\,dy,$$

where

$$W_1(\xi,y) = \frac{\partial W(\xi,y)}{\partial y} ,$$

hence

$$\left| \frac{W(\xi,t+h) - W(\xi,t)}{h} \right| \leq \psi(\xi) \in L_1(-\infty,\infty).$$

By Lebesgue's theorem on dominated convergence, we deduce that

$$\lim_{h\to 0} \int_{-\infty}^{\infty} \left| \frac{W(\xi,t+h) - W(\xi,t)}{h} - \frac{\partial W(\xi,t)}{\partial t} \right| d\xi = 0,$$

and (14.5) now yields U_4.

U_5. The equation

$$\frac{\partial^2 U(x,t)}{\partial x^2} = \frac{\partial U(x,t)}{\partial t}$$

holds for $t > 0$, $-\infty < x < \infty$, as can be verified directly by calculation.

Theorem 27. Given $f \in L_1(-\infty,\infty)$, *let* $U(x,t)$ *be any function with properties* U_1 *to* U_5. *Then*

$$U(x,t) = \int_{-\infty}^{\infty} f(\xi) W(x-\xi,t)\, d\xi, \quad t > 0,$$

where $W(\xi,t) = \frac{1}{2\sqrt{(\pi t)}} e^{-\xi^2/4t}$, *for almost all* $x \in (-\infty,\infty)$.

Proof. Let $\hat{U}(\alpha,t)$ be the Fourier transform of $U(x,t)$ considered as a function of x. It exists because of property U_1. By property U_2, together with (1.12), we have

(14.6) $$\hat{U}(\alpha,t) \to \hat{f}(\alpha), \text{ as } t \to 0+, \text{ for all } \alpha, \ -\infty < \alpha < \infty.$$

Because of property U_3, together with Theorem 3, we have

(14.7) $$\left[\frac{\partial^2 U}{\partial x^2}\right]^{\wedge}(\alpha) = (-i\alpha)^2\, \hat{U}(\alpha,t), \quad t > 0, \ -\infty < \alpha < \infty.$$

On the other hand,

$$\left[\frac{U(\cdot,t+h) - U(\cdot,t)}{h}\right]^{\wedge}(\alpha) = \frac{\hat{U}(\alpha,t+h) - \hat{U}(\alpha,t)}{h},$$

hence, by property U_4,

(14.8) $$\frac{\partial}{\partial t}[\hat{U}(\alpha,t)] = \left[\frac{\partial U(\cdot,t)}{\partial t}\right]^{\wedge}(\alpha).$$

But $$\frac{\partial^2 U(x,t)}{\partial x^2} = \frac{\partial U(x,t)}{\partial t}$$

because of property U_5, hence (14.7) and (14.8) yield

$$(-i\alpha)^2\ \hat{U}(\alpha,t) = \left[\frac{\partial^2 U(\cdot,t)}{\partial x^2}\right]^\wedge(\alpha) = \left[\frac{\partial U(\cdot,t)}{\partial t}\right]^\wedge(\alpha) = \frac{\partial}{\partial t}\ [\hat{U}(\alpha,t)].$$

Thus

$$\frac{\partial}{\partial t}\left[\hat{U}(\alpha,t)\right] = -\alpha^2\ \hat{U}(\alpha,t), \quad \text{for every real } \alpha,$$

hence

$$\hat{U}(\alpha,t) = A(\alpha)\ e^{-\alpha^2 t}, \quad \text{for every real } \alpha.$$

Since $\hat{U}(\alpha,t) \to \hat{f}(\alpha)$, as $t \to 0+$, by (14.6), we obtain

$$\hat{U}(\alpha,t) = \hat{f}(\alpha)\ e^{-\alpha^2 t}, \quad t > 0,\ -\infty < \alpha < \infty. \tag{14.9}$$

The Fourier transform of the convolution

$$(f*W)(x) = \int_{-\infty}^{\infty} f(\xi)W(x-\xi,t)d\xi, \quad t > 0,$$

however, equals $\hat{f}(\alpha)e^{-\alpha^2 t}$ (Theorem 2). Because of (14.9), and the uniqueness theorem (§6, Th.7), $U(\alpha,t) = (f*W)(\alpha)$, for $t > 0$, and for almost all $\alpha \in (-\infty,\infty)$.

The Cauchy-Poisson integral $V(f;x,t)$ of $f \in L_1(-\infty,\infty)$, given by (14.4) is associated in a similar fashion to the Laplace equation in (14.2). We list the characteristic properties of that integral:

V_1. For each $t > 0$, $V(x,t) \in L_1(-\infty,\infty)$ as a function of x, and

$$\|V(\cdot,t)\|_1 \le \|f\|_1.$$

V_2. We have

$$\lim_{t\to 0+} \|V(\cdot,t) - f(\cdot)\|_1 = 0,$$

as was proved in (9.6).

V_3. For $t > 0$, and $-\infty < x < \infty$, the partial derivatives $\frac{\partial V}{\partial x}$, $\frac{\partial^2 V}{\partial x^2}$ exist and belong to $L_1(-\infty,\infty)$ as functions of x.

V_4. For $t>0$, the partial derivatives $\frac{\partial V(x,t)}{\partial t}$, $\frac{\partial^2 V(x,t)}{\partial t^2}$ exist and belong to $L_1(-\infty,\infty)$ as functions of x, and

$$\lim_{h\to 0}\left\|\frac{V(\cdot,t+h)-V(\cdot,t)}{h}-\frac{\partial V(\cdot,t)}{\partial t}\right\|_1 = 0,$$

and

$$\lim_{h\to 0}\left\|\frac{\frac{\partial V(\cdot,t+h)}{\partial t}-\frac{\partial V(\cdot,t)}{\partial t}}{h}-\frac{\partial^2 V(\cdot,t)}{\partial t^2}\right\|_1 = 0.$$

V_5. The equation

$$\frac{\partial^2 V(x,t)}{\partial x^2}+\frac{\partial^2 V(x,t)}{\partial t^2} = 0$$

holds for all $x\in(-\infty,\infty)$ and $t>0$.

Theorem 28. Given $f\in L_1(-\infty,\infty)$, *let* $V(x,t)$ *be any function with properties* V_1 *to* V_5. *Then*

$$V(x,t) = \int_{-\infty}^{\infty} f(\xi)P(x-\xi,t)d\xi,\quad t>0,$$

where $P(\xi,t) = \frac{1}{\pi}\frac{t}{t^2+\xi^2}$, *for almost all* $x\in(-\infty,\infty)$.

Proof. Because of property V_1, the Fourier transform $\hat{V}(\alpha,t)$ is defined for each $t>0$, and $-\infty<\alpha<\infty$. By property V_3, and Theorem 3 of §1, we have

$$\left[\frac{\partial^2 V(\cdot,t)}{\partial x^2}\right]^{\wedge}(\alpha) = (-i\alpha)^2\,\hat{V}(\alpha,t); \tag{14.10}$$

on the other hand,

$$\left[\frac{V(\cdot,t+h)-V(\cdot,t)}{h}\right]^{\wedge}(\alpha) = \frac{\hat{V}(\alpha,t+h)-\hat{V}(\alpha,t)}{h},\quad h\neq 0,$$

and property V_4 implies that

$$\frac{\partial\hat{V}(\alpha,t)}{\partial t} = \left[\frac{\partial V(\cdot,t)}{\partial t}\right]^{\wedge}(\alpha),$$

and

$$\frac{\partial^2\hat{V}(\alpha,t)}{\partial t^2} = \left[\frac{\partial^2 V(\cdot,t)}{\partial t^2}\right]^{\wedge}(\alpha) = \left[-\frac{\partial^2 V(\cdot,t)}{\partial x^2}\right]^{\wedge}(\alpha)\qquad(\text{by } V_5)$$

$$= \alpha^2 \hat{V}(\alpha,t), \text{ by } (14.10).$$

Hence

(14.11) $$\hat{V}(\alpha,t) = A(\alpha)\, e^{-\alpha t} + B(\alpha)\, e^{\alpha t}, \text{ say.}$$

By property V_2, however, together with (1.12), we have

$$\hat{V}(\alpha,t) \to \hat{f}(\alpha), \text{ as } t \to 0+, \text{ for all } \alpha \in (-\infty,\infty),$$

hence

$$A(\alpha) + B(\alpha) = \hat{f}(\alpha).$$

By property V_1, we have, for each $t > 0$,

$$|\hat{V}(\alpha,t)| \leq \|f\|_1 < \infty,$$

which, in turn, implies that

$$|B(\alpha)e^{\alpha t}| \leq \|f\|_1 + |A(\alpha)e^{-\alpha t}| \leq \|f\|_1 + |A(\alpha)|,$$

because of (14.11). Hence

$$|B(\alpha)| \leq \frac{\|f\|_1 + |A(\alpha)|}{|e^{\alpha t}|},$$

and on letting $t \to \infty$, we get $B(\alpha) = 0$, for $\alpha > 0$; similarly also $A(\alpha) = 0$ for $\alpha < 0$; so that $\hat{V}(\alpha,t) = \hat{f}(\alpha)e^{-|\alpha|t}$, for $\alpha \in (-\infty,\infty)$. But the Fourier transform of the convolution

$$(f*P)(x) = \int_{-\infty}^{\infty} f(\xi)P(x-\xi,t)d\xi, \quad t > 0,$$

equals $\hat{f}(\alpha)e^{-|\alpha|t}$ (by Theorem 2, §1). Because of the uniqueness Theorem (Th.7, §6) it follows that $V(x,t) = (f*P)(x)$ for almost all $x \in (-\infty,\infty)$, and $t > 0$.

§15. Several variables

The definition of Fourier transform is easily extended to functions of several variables. If E_k denotes the real Euclidean space of dimension k, and $x \in E_k$, we write $x = (x_1,\dots,x_k)$, where $-\infty < x_r < \infty$, for $r = 1,2,\dots,k$; we write $|x| = (x_1^2 + \dots + x_k^2)^{1/2}$, and if $\alpha \in E_k$, we write $\langle x,\alpha\rangle = x_1\alpha_1 + \dots + x_k\alpha_k$.

For any p such that $1 \le p < \infty$, we denote by $L_p(E_k)$ the Banach space of complex-valued, Lebesgue measurable functions f on E_k, relative to the norm

$$\|f\|_p = \left(\int_{E_k} |f(x)|^p dx\right)^{1/p} < \infty,$$

modulo the subspace of functions which are zero almost everywhere. Here dx ($= dx_1 \dots dx_k$) stands for the k-dimensional Lebesgue measure.

If $f \in L_1(E_k)$, we define $\hat{f}$, the Fourier transform of f, by the relation

$$(15.1)\qquad \hat{f}(\alpha) = \int_{E_k} f(x)\, e^{i\langle x,\alpha\rangle} dx, \quad \text{with } F[f](\alpha) = (2\pi)^{-k/2}\hat{f}(\alpha),$$

for all $\alpha \in E_k$. We note that $\hat{f}$ is bounded, and continuous, and

$$(15.2)\qquad |\hat{f}(\alpha)| \le \|f\|_1 < \infty,$$

as in (1.5) and (1.6).

If τ_h is the operator which takes f(x) into f(x+h), where $x,h \in E_k$, then

$$(15.3)\qquad (\tau_h f)^\wedge(x) = e^{-i\langle h,x\rangle}\hat{f}(x).$$

The Fourier transform of $f(\lambda x_1,\lambda x_2,\dots,\lambda x_k)$, where λ is a real number, and $\lambda \neq 0$, is $\frac{1}{|\lambda|^k}$ times the Fourier transform $\hat{f}\left(\frac{x_1}{\lambda},\dots,\frac{x_k}{\lambda}\right)$, as in (1.10).

The Riemann-Lebesgue theorem holds: if $f \in L_1(E_k)$, then $\hat{f}(x) \to 0$, as $|x| \to \infty$. The proof is similar to that of Theorem 1. We approximate to f in the L_1-norm by box-functions in E_k. A box-function g is such that $g(x) = 1$ in the box $-\infty < a_r \le x_r \le b_r < \infty$, for $r = 1,2,\dots,k$, and $g(x) = 0$ outside the box.

As in (1.13) the composition rule holds: if $f,g \in L_1(E_k)$, then

$$(15.4) \qquad \int_{E_k} f(x)\hat{g}(x)dx = \int_{E_k} \hat{f}(y)g(y)dy.$$

As in §2, one defines the *convolution* of $f \in L_1(E_k)$ and $g \in L_1(E_k)$, written $h = f*g$, by the relation

$$(15.5) \qquad h(x) = \int_{E_k} f(x-y)g(y)dy,$$

and notes that $h \in L_1(E_k)$. The operation of convolution is commutative and associative, and

$$(15.6) \qquad (f*g)^\wedge = \hat{f}\cdot\hat{g}$$

as in Theorem 2.

As for pointwise summability, let $K \in L_1(E_k)$, and $\hat{K} = H$. For $R > 0$, let $K_R(x) = K(\frac{x}{R})$, and $H_R(x) = R^k H(Rx)$, as in (7.1). Then for every $f \in L_1(E_k)$, we have, by the composition rule (15.4),

$$(15.7) \qquad \int_{E_k} \hat{f}(\alpha)e^{-i<\alpha,x>}K(\tfrac{\alpha}{R})d\alpha = \int_{E_k} f(x+u)H_R(u)du = \int_{E_k} f(y)H_R(y-x)dy.$$

If we take, *in particular*,

$$(15.8) \qquad K(x) = e^{-|x|^2}, \quad |x|^2 = x_1^2 + \dots + x_k^2,$$

then K is a *radial* function; that is to say, $K(x) = K(y)$ if $|x| = |y|$. If $k = 1$, a radial function is just an even function. By direct calculation (as a product of one-dimensional Fourier transforms, using Ex.6, §1), we see that

$$(15.9) \qquad H_R(x) \equiv R^k H(Rx) = \hat{K}_R(x) = \pi^{k/2}\cdot R^k\cdot e^{-|x|^2R^2/4}$$

is also *radial* , so that $H_R(y-x) = H_R(x-y)$, and (15.7) leads, for the particular choice of K in (15.8), to the relation

$$(15.10) \qquad (f*H_R)(x) = \int_{E_k} \hat{f}(\alpha)e^{-i<\alpha,x>}\cdot K(\tfrac{\alpha}{R})d\alpha, \quad R > 0,\ K(\alpha) = e^{-|\alpha|^2},$$

which is analogous to (7.1). We note further that

$$(15.11) \qquad (2\pi)^{-k} \int_{E_k} H_R(x)dx = 1$$

where H_R is defined as in (15.9), if we use Example 6, §1. Hence (15.10) implies that

$$(15.12)\qquad (2\pi)^{-k}(f*H_R)(x) - f(x) = (2\pi)^{-k} \int_{E_k} \{f(x+u) - f(x)\}H_R(u)du,$$

where $H_R(u)$ is a radial function, which depends only on $|u|$. Integrating first over the "surface" $|u| = t > 0$, and then relative to the "radius" t, for $0 \leq t < \infty$, we obtain

$$(15.3)\qquad (2\pi)^{-k}(f*H_R)(x) - f(x)$$

$$= (2\pi)^{-k} \int_0^\infty t^{k-1} \left(\int_\sigma \{f(x+tv) - f(x)\}d\sigma_v\right) H_R(t)dt,$$

where σ denotes the "sphere": $v_1^2 + \ldots + v_k^2 = 1$, and $d\sigma_v$ its (k-1)-dimensional volume element.

We now define for $x \in E_k$, $k \geq 2$, and $t > 0$,

$$(15.14)\qquad f_x(t) = \frac{1}{\omega_{k-1}} \int_\sigma f(x+tv)d\sigma_v,$$

where σ denotes the "sphere" $v_1^2 + \ldots v_k^2 = 1$, ω_{k-1} its (k-1)dimensional volume, and $d\sigma_v$ its (k-1)-dimensional volume element.

For k = 1, we define $f_x(t) = \frac{1}{2}[f(x+t) + f(x-t)]$. If

$$(15.15)\qquad g_x(t) = t^{k-1}[f_x(t) - f(x)],\ 0 \leq t < \infty,$$

then relation (15.13) can be written as

$$(15.16)\qquad (2\pi)^{-k}(f*H_R)(x) - f(x) = \omega_{k-1}(2\pi)^{-k} \int_0^\infty R^k H(Rt)g_x(t)dt,$$

which is the analogue of (7.2) in Theorem 8.

Following the same lines of proof as in Theorem 9, we can deduce that

$$(15.17)\qquad \lim_{R\to\infty} \frac{1}{(2\pi)^k} (f*H_R)(x) = f(x)\ ,$$

at every point x at which

$$\lim_{T\to 0} \frac{1}{T^k} \int_0^T g_x(t)\,dt = 0. \tag{15.18}$$

By (15.10) it follows that if $f \in L_1(E_k)$, then

$$\lim_{R\to\infty} \int_{E_k} \hat{f}(\alpha)\, e^{-i<\alpha,x>}\, e^{-|\alpha|^2/R^2}\, d\alpha = f(x) \tag{15.19}$$

for almost every $x \in E_k$, since, by a Theorem of Lebesgue-Vitali, condition (15.18) is satisfied for almost every x, if $f \in L_1(E_k)$. Thus the Fourier integral of any function in $L_1(E_k)$ is Gauss-summable almost everywhere (cf. (7.13)).

This can again be used, as in Theorem 11, to prove that if $f \in L_1(E_k)$ and $\hat{f} \in L_1(E_k)$, then the inversion formula holds almost everywhere, that is to say:

$$f(x) = \frac{1}{(2\pi)^k} \int_{E_k} \hat{f}(\alpha) e^{-i<\alpha,x>} d\alpha \tag{15.20}$$

for almost every $x \in E_k$.

Corresponding to Corollary (8.7), we obtain the result that if $f \in L_1(E_k)$, is non-negative, and continuous at the origin, then $\hat{f} \in L_1(E_k)$.

As for summability in the L_1-norm, one can prove, using (15.12), the analogue of Theorem 15, in an analogous manner: we have

$$\|(2\pi)^{-k}(f*H_R) - f\|_1 \to 0, \text{ as } R\to\infty, \tag{15.21}$$

for the particular kernel (15.9).

Instead of choosing K to be the Gauss kernel $e^{-|x|^2}$ as in (15.8), one can choose K to be any radial function, such that $K \in L_1(E_k)$, $K(0)= 1$, K is continuous at the origin, and $(2\pi)^{-k} \int_{-\infty}^{\infty} \hat{K}(\alpha)\,d\alpha = 1$, and prove the analogues of (15.7) and (15.10). We shall not elaborate here on *general* summability.

On the other hand, it is worth noting that if $f \in L_1(E_k)$, *and* f is radial, then $\hat{f}$ is also radial, and assumes a special form depending on whether k is even or odd.

For we have

$$\hat{f}(\alpha) = \int_{E_k} f(|x|)e^{i<\alpha,x>}dx, \quad \alpha \in E_k, \tag{15.22}$$

and if T: $x \to y$ is any orthogonal transformation of E_k, represented by the matrix (b_{ij}) with determinant +1, such that $y_i = \sum_{j=1}^{k} b_{ij}x_j$, $i = 1,\ldots,k$, we can choose the first row of the matrix $b_{1j} = \frac{\alpha_j}{|\alpha|}$ for $j = 1,\ldots,k$. Then we have $<\alpha,x> \equiv \sum_{i=1}^{k} \alpha_i x_i = |\alpha|\cdot y_1$, while $|x| = |y|$, since T is orthogonal. Hence

$$\hat{f}(\alpha) = \int_{E_k} f(|y|)e^{i|\alpha|y_1}dy, \quad \alpha \in E_k, \ y = (y_1,\ldots,y_k),$$

which shows that $\hat{f}(\alpha)$ depends on $|\alpha|$ alone, hence is *radial*.

Let $\rho = (y_2^2 + \ldots + y_k^2)^{1/2}$, so that $\rho \geq 0$, and $|y| = (y_1^2 + \rho^2)^{1/2}$. We integrate first over the "surface" of the (k-1) dimensional sphere $y_2^2 + \ldots + y_k^2 = \rho^2$, and then relative to the "radius" ρ from 0 to ∞. Thus

$$\hat{f}(\alpha) = \int_{-\infty}^{\infty} e^{i|\alpha|y_1}dy_1 \int_0^{\infty} f\left\{(y_1^2 + \rho^2)^{1/2}\right\} \rho^{k-2} \omega_{k-2}\, d\rho,$$

where ω_{k-2} stands for the (k-2)-dimensional volume of the "sphere" $y_2^2 + \ldots + y_k^2 = 1$. If we now use polar coordinates: $y_1 = r\cos\theta$, $\rho = r\sin\theta$, and note that $0 \leq \theta \leq \pi$, since $\rho \geq 0$, while $r \geq 0$, we obtain

$$\hat{f}(\alpha) = \omega_{k-2} \int_0^{\infty}\int_0^{\pi} f(r)r^{k-1} e^{i|\alpha|r\cos\theta}(\sin\theta)^{k-2}dr\,d\theta.$$

From Example 8, of §1, we have

$$\int_0^{\pi} e^{i|\alpha|\cos\theta}(\sin\theta)^{2\nu}d\theta = 2^{\nu}\Gamma(\nu+\tfrac{1}{2})\Gamma(\tfrac{1}{2})\ J_{\nu}(|\alpha|)\ |\alpha|^{-\nu}, \quad \nu > -\tfrac{1}{2}, \tag{15.23}$$

where J_ν is the Bessel function of order ν, hence

$$\hat{f}(\alpha) = \frac{c\cdot\omega_{k-2}}{|\alpha|^{(k-2)/2}} \int_0^{\infty} f(r)r^{k/2}J_{\frac{1}{2}(k-2)}(|\alpha|r)dr,$$

$$c = 2^{(k-2)/2}\Gamma(\tfrac{k-1}{2})\Gamma(\tfrac{1}{2}),$$

for $k \geq 2$, with $\omega_0 \equiv 2$.

Since the volume $V_k(s)$ of the (solid) sphere defined by $|y|^2 \leq s$ equals $\left\{\pi^{k/2}s^{k/2}/\Gamma(\frac{1}{2}k+1)\right\}$, which can be proved by induction on k, we have $\omega_{k-1} = \left\{2\pi^{k/2}/\Gamma(k/2)\right\}$, for $k \geq 2$. We thus have for any $f \in L_1(E_k)$, where f is *radial*,

$$\hat{f}(\alpha) = \frac{(2\pi)^{k/2}}{|\alpha|^{(k-2)/2}} \int_0^\infty f(r)r^{k/2}J_{\frac{1}{2}(k-2)}(|\alpha|r)dr. \tag{15.24}$$

If $k = 3$, then $(k-2)/2 = \frac{1}{2}$, and $J_{\frac{1}{2}}(x) = \left(\frac{2}{\pi x}\right)^{1/2} \sin x$, because of (15.23) with $\nu = \frac{1}{2}$, hence

$$\hat{f}(\alpha) = \frac{4\pi}{|\alpha|} \int_0^\infty rf(r)\sin(|\alpha|r)dr.$$

If $k = 2$, then obviously

$$\hat{f}(\alpha) = 2\pi \int_0^\infty rf(r)J_0(|\alpha|r)dr.$$

Formula (15.24) holds good also for $k = 1$, since $J_{-\frac{1}{2}}(x) = \left(\frac{2}{\pi x}\right)^{1/2}\cos x$, as can be seen directly by taking $\nu = -\frac{1}{2}$ in the defining series for J_ν given in Example 8, §1.

In general, a Bessel function appears in the integral for $\hat{f}$ if k is *even* , and a trigonometric function if k is odd.

Chapter II. Fourier transforms on $L_2(-\infty,\infty)$

§1. Introduction

The Banach space $L_2(-\infty,\infty)$ is endowed with an *inner product*. For any two functions f,g belonging to $L_2(-\infty,\infty)$, the inner product (f,g) is defined by

(1.1) $$(f,g) = \int_{-\infty}^{\infty} f(x)\overline{g}(x)\,dx,$$

the integral existing because of Schwarz's inequality. (The bar "—" denotes complex conjugation). It has the following properties:
(i) $\overline{(f,g)} = (g,f)$, (ii) $(f,f) = \|f\|_2^2$, (iii) $(f,ag) = \overline{a}(f,g)$, $a \in \mathbb{C}$, (iv) $(f,g_1+g_2) = (f,g_1) + (f,g_2)$, for $f,g_1,g_2 \in L_2(-\infty,\infty)$, (v) $|(f,g)| \leq \|f\|_2 \cdot \|g\|_2$, (vi) (f,g) is continuous in f, and in g.

A Banach space with such an inner product is referred to as a Hilbert space, and the particular Hilbert space $L_2(-\infty,\infty)$ is *separable*, in the sense that it contains a *countable* subset which is *dense* in $L_2(-\infty,\infty)$.

If a function f belongs to $L_2(-\infty,\infty)$, it does not necessarily follow that $f \in L_1(-\infty,\infty)$, as for example $f(x) = \frac{1}{1+|x|}$. One has therefore to define the Fourier transform on $L_2(-\infty,\infty)$ somewhat differently from the way it was done in the case of $L_1(-\infty,\infty)$ in Chapter I, but make sure that the two definitions coincide if $f \in L_1(-\infty,\infty) \cap L_2(-\infty,\infty)$.

We shall define a Fourier transform first on a *dense subset* of $L_2(-\infty,\infty)$, and then extend it uniquely to the whole of $L_2(-\infty,\infty)$. The extension will be defined *almost* everywhere, instead of everywhere as in the case of $L_1(-\infty,\infty)$. One can start with the dense subset of bounded functions belonging to $L_1(-\infty,\infty)$, or the subset $L_1(-\infty,\infty) \cap L_2(-\infty,\infty)$, or the subset of continuous functions which are of bounded variation over a finite interval and vanish outside that interval, or the subset of step-functions, or the subset S of infinitely differentiable functions which "decrease rapidly" (Schwartz's space, §3, Ch.I). For convenience we shall choose the last alternative.

The Fourier transform will turn out to be a bounded, linear operator on S, which is an isometry, and which can be uniquely extended to a bounded linear operator on all of $L_2(-\infty,\infty)$, whose range coincides with $L_2(-\infty,\infty)$, and which is "isometric", hence a *unitary operator*. Thus if $f \in L_2(-\infty,\infty)$, its Fourier transform $F[f]$ automatically belongs to $L_2(-\infty,\infty)$. Because of the symmetry typified by this fact, we define $F[f] = \frac{1}{\sqrt{(2\pi)}}\hat{f}$, for $f \in L_1(-\infty,\infty) \cap L_2(-\infty,\infty)$ (cf. (1.2) of Ch.I).

§2. Plancherel's theorem

Fourier transforms on S. If S denotes Schwartz's space (as in §3, Ch.I), and $f \in S$, we define the Fourier transform $F[f]$ of f by the relation

$$(2.1) \qquad F[f] = \frac{1}{\sqrt{(2\pi)}}\hat{f}, \text{ or}$$

$$F[f](\alpha) = \frac{1}{\sqrt{(2\pi)}}\hat{f}(\alpha) = \frac{1}{\sqrt{(2\pi)}}\int_{-\infty}^{\infty} f(x)e^{i\alpha x}dx, \qquad -\infty < \alpha < \infty.$$

The map $f \to F[f]$ is *a linear map of S onto itself*, from what we have already seen (in §8, Ch.I). If we define $\check{F}[f]$ by the relation

$$(2.2) \qquad \check{F}[f](\alpha) = F[f](-\alpha) = \frac{1}{\sqrt{(2\pi)}}\hat{f}(-\alpha)$$

$$= \frac{1}{\sqrt{(2\pi)}}\int_{-\infty}^{\infty} f(x)e^{-i\alpha x}dx, \qquad -\infty < \alpha < \infty,$$

then (by 8.12), Ch.I, p. 50) we have

$$(2.3) \qquad \check{F}[Ff] = f.$$

Further if $f,g \in S$, then (by (8.13), Ch.I, p. 50) we have (cf. (1.1))

$$(2.4) \qquad (f,g) = (F[f], F[g]); \quad \|f\|_2 = \|F[f]\|_2;$$

and

(2.5) $$\frac{1}{\sqrt{(2\pi)}} F[f*g] = F[f]\cdot F[g],$$

where the star "*" denotes convolution (see (8.14), Ch.I, p. 50).

Fourier transforms on $L_2(-\infty,\infty)$

Since S is a dense subset of $L_2(-\infty,\infty)$, if $f \in L_2(-\infty,\infty)$, there exists a sequence (f_n) of functions belonging to S, such that $\|f-f_n\|_2 \to 0$, as $n \to \infty$. Hence $\|f_m-f_n\|_2 \to 0$, as $m,n \to \infty$, but by (2.4), and the linearity of the map $f_m \to F[f_m]$, $\|F[f_m] - F[f_n]\|_2 \to 0$, as $m,n \to \infty$. Since the *space* $L_2(-\infty,\infty)$ is *complete*, there exists a function $F \in L_2(-\infty,\infty)$, such that $\|F[f_n] - F\|_2 \to 0$, as $n \to \infty$. The function F is defined *almost* everywhere on $(-\infty,\infty)$. We call F the *Fourier transform* of $f \in L_2(-\infty,\infty)$, and *denote it* by $F[f]$.

We note that F does not depend on the approximating sequence (f_n). Let (g_n) be another sequence of functions belonging to S, such that $\|g_n-f\|_2 \to 0$. Then $f_n-g_n = f_n-f + f-g_n$, and $\|f_n-g_n\|_2 \le \|f_n-f\|_2 + \|g_n-f\|_2 \to 0$, as $n \to \infty$. But, by (2.4), $\|F[f_n] - F[g_n]\|_2 = \|f_n-g_n\|_2 \to 0$, hence $\|F[g_n] - F\|_2 \le \|F - F[f_n]\|_2 + \|F[f_n] - F[g_n]\|_2 \to 0$, as $n \to \infty$.

We are thus led to the following

(2.6) Definition. If $f \in L_2(-\infty,\infty)$, the Fourier transform $F[f]$ of f is defined to be the limit, in the L_2-norm, of the sequence $\{F[f_n]\}$ of Fourier transforms, of *any* sequence (f_n) of functions belonging to S, such that f_n converges in the L_2-norm to the given function $f \in L_2(-\infty,\infty)$, as $n \to \infty$. The function $F[f]$ is defined almost everywhere on $(-\infty,\infty)$, and belongs to $L_2(-\infty,\infty)$.

The definition makes it clear that the map $f \to F[f]$ of $L_2(-\infty,\infty)$ *into* $L_2(-\infty,\infty)$ is *linear*. That is to say, if $f_1,f_2 \in L_2(-\infty,\infty)$, and $a,b \in \mathbb{C}$, then

(2.7) $$F[af_1 + bf_2] = aF[f_1] + bF[f_2].$$

We note further that for *any* $f \in L_2(-\infty,\infty)$, we have

(2.8) $$\|f\|_2 = \|F[f]\|_2,$$

since, by definition (2.6), and (2.4), we have

(2.9) $$\|f\|_2 = \lim_{n\to\infty} \|f_n\|_2 = \lim_{n\to\infty} \|F[f_n]\|_2 = \|F[f]\|_2.$$

If $f_1, f_2 \in L_2(-\infty,\infty)$, then we can apply (2.8) to the function $f = f_1 + f_2$, and noting that $\|f_1\|_2 = \|F[f_1]\|_2$, $\|f_2\|_2 = \|F[f_2]\|_2$, deduce that

$$\mathrm{Re}\left(\int_{-\infty}^{\infty} f_1(x)\overline{f_2}(x)\,dx\right) = \mathrm{Re}\left(\int_{-\infty}^{\infty} F[f_1](x)\cdot\overline{F[f_2]}(x)\,dx\right).$$

Replacing f_1 by if_1, we see that

$$\mathrm{Im}\left(\int_{-\infty}^{\infty} f_1(x)\ \overline{f_2}(x)\,dx\right) = \mathrm{Im}\left(\int_{-\infty}^{\infty} F[f_1](x)\cdot\overline{F[f_2]}(x)\,dx\right).$$

Hence we see that (2.8) implies, in the notation of (1.1), that

(2.10) $$(f_1,f_2) = (F[f_1],\ F[f_2]), \quad \text{for } f_1,f_2 \in L_2(-\infty,\infty).$$

We can similarly define $\check{F}[f]$ for *any* $f \in L_2(-\infty,\infty)$ by extending definition (2.2) from S to $L_2(-\infty,\infty)$. Thus given any $f \in L_2(-\infty,\infty)$, there exists a sequence (f_n), $f_n \in S$ for $n = 1,2,\ldots$, such that $\|f-f_n\|_2 \to 0$, as $n\to\infty$. The sequence $(\check{F}[f_n])$, where $\check{F}[f_n] \in S$, converges in the L_2-norm to a function in $L_2(-\infty,\infty)$, which we *define* to be $\check{F}[f]$. It is defined almost everywhere on $(-\infty,\infty)$, and we have

$$\check{F}[f](\alpha) = F[f](-\alpha)$$

for almost all $\alpha \in (-\infty,\infty)$.

The "inversion formula" holds almost everywhere, without any special artifice, since we have

(2.11) $$\check{F}[F[f]] = f = F[\check{F}[f]], \text{ for } f \in L_2(-\infty,\infty).$$

For if $f \in L_2(-\infty,\infty)$, and $f_n \in S$, $n = 1,2,\ldots$, and $\|f-f_n\|_2 \to 0$, as $n\to\infty$, we have, by definition, $\lim_{n\to\infty} \|F[f_n] - F[f]\|_2 = 0$, $\lim_{n\to\infty} \|\check{F}[f_n] - \check{F}[f]\|_2 = 0$, where $F[f]$, $\check{F}[f] \in L_2(-\infty,\infty)$, and since (2.11) holds for any $f \in S$ (cf. (2.3)), we see that

$$\|f-\check{F}[F[f]]\|_2 \le \|f-f_n\|_2+\|f_n-\check{F}[F[f_n]]\|_2+\|\check{F}[F[f_n]]-\check{F}[F[f]]\|_2 \to 0,$$

as $n\to\infty$, hence $f = \check{F}[F[f]]$ almost everywhere, and similarly also $F[\check{F}[f]] = f$, giving (2.11), which shows that the map $f \to F[f]$ is a linear map of $L_2(-\infty,\infty)$ *onto* itself.

We shall now show that $F[f]$ thus defined for $f \in L_2(-\infty,\infty)$ is related to the integral defining $\frac{1}{\sqrt{(2\pi)}}\hat{f}$ in the case of $f \in L_1(-\infty,\infty)$.

Let $f \in L_2(-\infty,\infty)$, and let $f(x) = 0$, for $|x| \geq A > 0$. Then we shall see that

$$(2.12) \qquad F[f](\alpha) = \frac{\hat{f}(\alpha)}{\sqrt{(2\pi)}} = \frac{1}{\sqrt{(2\pi)}} \int_{-\infty}^{\infty} f(x)e^{i\alpha x}dx, \text{ for almost all } \alpha \in (-\infty,\infty).$$

If we *denote* the integral on the right-hand side of (2.12) by $f^*(\alpha)$, we shall show that $F[f](\alpha) = f^*(\alpha)$ for almost all $\alpha \in (-\infty,\infty)$.

There exists a sequence (f_n), $n = 1,2,\ldots$, with $f_n \in S$, such that every member of the sequence vanishes outside the interval $(-A,A)$, and $\|f-f_n\|_2 \to 0$, as $n \to \infty$. That implies, by (2.7) and (2.8), that $\|F[f] - F[f_n]\|_2 \to 0$, and, in particular,

$$(2.13) \qquad \int_{-R}^{R} |F[f_n](x) - F[f](x)|^2 dx \to 0, \text{ as } n \to \infty,$$

for each $R > 0$. But

$$|F[f_n](\alpha)-f^*(\alpha)| \leq \int_{-A}^{A} |f_n(x)-f(x)|dx \leq \left(2A \int_{-A}^{A} |f_n(x)-f(x)|^2 dx\right)^{1/2} \to 0, \text{ as } n \to \infty,$$

hence $F[f_n] \to f^*$ *uniformly* over *every finite interval*, and, in particular,

$$(2.14) \qquad \int_{-R}^{R} |F[f_n](x) - f^*(x)|^2 dx \to 0, \text{ as } n \to \infty.$$

A comparison of (2.13) and (2.14) shows that $f^* = F[f]$ almost everywhere on $(-R,R)$, for *each* $R > 0$, and hence almost everywhere on $(-\infty,\infty)$, thus proving (2.12).

Finally let $f \in L_2(-\infty,\infty)$ without any further condition. We define the function f_R, for each $R > 0$, by the requirement

$$f_R(x) = \begin{cases} f(x), & \text{for } |x| < R, \\ 0, & \text{for } |x| \geq R. \end{cases}$$

Then $f_R \in L_1(-\infty,\infty) \cap L_2(-\infty,\infty)$, and by what has just been proved in (2.12), we have

(2.15) $$F[f_R](\alpha) = \frac{1}{\sqrt{(2\pi)}} \int_{-R}^{R} f(x)e^{i\alpha x}dx, \quad -\infty < \alpha < \infty.$$

On the other hand, since $f \to F[f]$ is a linear map of $L_2(-\infty,\infty)$ onto itself, we have, by (2.8),

(2.16) $$\|F[f]-F[f_R]\|_2 = \|F[f-f_R]\|_2 = \|f-f_R\|_2 \to 0, \text{ as } R \to \infty.$$

Hence for each $f \in L_2(-\infty,\infty)$, the integral $F[f_R]$ given by (2.15) converges in the L_2-norm to $F[f] \in L_2(-\infty,\infty)$, as $R \to \infty$. This implies, by Weyl's formulation of the Riesz-Fischer theorem, that there exists a sequence (R_k), with $R_k > 0$ for $k = 1,2,...$, such that

$$\frac{1}{\sqrt{(2\pi)}} \int_{-R_k}^{R_k} f(x)e^{ix\alpha}dx \to F[f](\alpha),$$

as $R_k \to \infty$, for almost every $\alpha \in (-\infty,\infty)$. In particular, if the integral

(2.17) $$\frac{1}{\sqrt{(2\pi)}} \int_{-\infty}^{\infty} f(x)e^{ix\alpha}dx$$

exists as a Cauchy principal value for almost all $\alpha \in (-\infty,\infty)$, it equals $F[f]$. And if $f \in L_1(-\infty,\infty) \cap L_2(-\infty,\infty)$, then

(2.18) $$F[f](\alpha) = \frac{1}{\sqrt{(2\pi)}} \int_{-\infty}^{\infty} f(x)e^{i\alpha x}dx, \quad -\infty < \alpha < \infty.$$

We subsume the results of (2.6) - (2.11), (2.17) and (2.18) under the following

Theorem 1 (Plancherel). If $f \in L_2(-\infty,\infty)$, *then there exists a function* $F[f] \in L_2(-\infty,\infty)$, *designated the Fourier transform of* f, *such that, for any real* α,

(2.19) $$\frac{1}{\sqrt{(2\pi)}} \int_{-R}^{R} f(x)e^{i\alpha x}dx \to F[f](\alpha), \text{ in the } L_2\text{-norm, as } R \to \infty,$$

and

(2.20) $$\frac{1}{\sqrt{(2\pi)}} \int_{-R}^{R} F[f](\alpha)\cdot e^{-i\alpha x}d\alpha \to f(x), \text{ in the } L_2\text{-norm, as } R \to \infty,$$

with

(2.21) $$\|f\|_2 = \|F[f]\|_2.$$

Every function $f \in L_2(-\infty,\infty)$ *is the Fourier transform of a unique element of* $L_2(-\infty,\infty)$.

As in the L_1-case (see Ch.I, (1.8), (1.9)), we have for $f \in L_2(-\infty,\infty)$, and any real number a, the relations

(2.22) $$F[f(\cdot+a)](y) = e^{-iya}F[f](y) \; ; \quad F[\overline{f}(\cdot)](y) = \overline{F[f]}(-y);$$

while

(2.23) $$\check{F}[f(\cdot+a)](y) = e^{iya}F[f](-y) \; ; \quad \check{F}[\overline{f}(\cdot)](y) = \overline{F[f]}(y).$$

To indicate the reasoning involved, let us consider the first relation in (2.22). It follows from the fact that

$$\int_{-R}^{R} f(x+a)e^{iyx}dx = \int_{-R+a}^{R+a} f(x)e^{iy(x-a)}dx = e^{-iya}\int_{-R+a}^{R+a} f(x)e^{iyx}dx,$$

where

$$\int_{-R+a}^{-R} f(x)e^{iyx}dx \to 0, \quad \text{and} \int_{R}^{R+a} f(x)e^{iyx}dx \to 0,$$

in the L_2-norm, as $R \to \infty$, because of (2.9).

For $f,g \in L_2(-\infty,\infty)$, we have already shown in (2.10) that (2.8) implies that

(2.24) $$\int_{-\infty}^{\infty} f(x)\overline{g}(x)dx = \int_{-\infty}^{\infty} F[f](x)\; \overline{F[g]}(x)dx.$$

On using (2.22) and (2.23) in (2.24), we get the following relations:

(2.25) $$\int_{-\infty}^{\infty} f(x)g(-x)dx = \int_{-\infty}^{\infty} F[f](y)\; F[g](y)dy$$

(2.26) $$\int_{-\infty}^{\infty} f(x)g(a-x)dx = \int_{-\infty}^{\infty} F[f](x)\; F[g](x)e^{-iax}dx$$

and

(2.27) $$\int_{-\infty}^{\infty} f(t)g(t)e^{ixt}dt = \int_{-\infty}^{\infty} F[f](t)\; F[g](x-t)dt,$$

all the integrals being absolutely convergent.

We may look upon (2.26) as the L_2-analogue of formula (2.2) of Chapter I.

If we choose

$$g(x) = g_{a,b}(x) = \begin{cases} 1, & \text{for } a < x < b, \\ 0, & \text{for } x \leq a,\ x \geq b, \end{cases}$$

then $g \in L_1(-\infty,\infty) \cdot\cap\cdot L_2(-\infty,\infty)$, and

$$F[g](y) = \frac{1}{\sqrt{(2\pi)}} \int_a^b e^{ixy}dx = \frac{1}{\sqrt{(2\pi)}} \frac{e^{iby}-e^{iay}}{iy} \in L_2(-\infty < y < \infty),$$

and (2.27), with $x = 0$, gives

$$\text{(2.28)} \qquad \int_a^b f(t)dt = \frac{1}{\sqrt{(2\pi)}} \int_{-\infty}^{\infty} F[f](y) \frac{e^{-iby}-e^{-iay}}{-iy} dy, \quad a < b.$$

On the other hand, we may take

$$F[g](y) = F[g_{a,b}](y) = \begin{cases} 1, & \text{for } a < y < b, \\ 0, & \text{for } y \leq a,\ y \geq b. \end{cases}$$

Then

$$g(x) = \frac{1}{\sqrt{(2\pi)}} \int_{-\infty}^{\infty} F[g_{a,b}](y)\, e^{-ixy}dy = \frac{1}{\sqrt{(2\pi)}} \int_a^b e^{-ixy}dy$$

$$= \frac{1}{\sqrt{(2\pi)}} \frac{e^{-ixb}-e^{-ixa}}{-ix} \in L_2(-\infty < x < \infty),$$

and (2.25) gives

$$\text{(2.29)} \qquad \int_a^b F[f](x)dx = \frac{1}{\sqrt{(2\pi)}} \int_{-\infty}^{\infty} f(x) \frac{e^{ibx}-e^{iax}}{ix} dx, \quad a < b.$$

Remarks. The operator F has been defined on $L_1(-\infty,\infty)$ in Chapter I, (1.2), and on $L_2(-\infty,\infty)$ in Plancherel's theorem. The definition can be extended to the space $L_1(-\infty,\infty) + L_2(-\infty,\infty)$ consisting of all functions f of the form $f = f_1 + f_2$, where $f_1 \in L_1(-\infty,\infty)$, $f_2 \in L_2(-\infty,\infty)$.

For such an f we *define* $F[f] = F[f_1] + F[f_2]$. This definition does not depend on the particular decomposition $f = f_1 + f_2$. For if $f = g_1 + g_2$, $g_1 \in L_1(-\infty,\infty)$, $g_2 \in L_2(-\infty,\infty)$ is a different decomposition, then $f_1 - g_1 = g_2 - f_2 \in L_1(-\infty,\infty) \cap L_2(-\infty,\infty)$ on which, by (2.18), the L_1-definition and the L_2-definition of F coincide, so that $F[f_1 - g_1] = F[f_1] - F[g_1] = F[g_2] - F[f_2]$, and we have $F[f_1] + F[f_2] = F[g_1] + F[g_2]$. Thus the operator F is defined on the space $L_1(-\infty,\infty) + L_2(-\infty,\infty)$, which, in fact, contains all the spaces $L_p(-\infty,\infty)$, for $1 \le p \le 2$. For if $g \in L_p(-\infty,\infty)$, $1 < p < 2$, then $g = g_1 + g_2$, where $g_1 \in L_1(-\infty,\infty)$, and $g_2 \in L_2(-\infty,\infty)$. We have only to define

$$g_1(x) = \begin{cases} g(x), & \text{if } |g(x)| \ge 1, \\ 0, & \text{if } |g(x)| < 1, \end{cases} \quad \text{and} \quad g_2(x) = \begin{cases} g(x), & \text{if } |g(x)| < 1, \\ 0, & \text{if } |g(x)| \ge 1, \end{cases}$$

so that

$$\int_{-\infty}^{\infty} |g_1(x)|\,dx = \int_{[x|\,|g(x)|\ge 1]} |g(x)|\,dx$$

$$\le \int_{[x|\,|g(x)|\ge 1]} |g(x)|^{1+(p-1)}\,dx \le \int_{-\infty}^{\infty} |g(x)|^p\,dx < \infty,$$

while g_2 is bounded, and belongs to $L_p(-\infty,\infty)$, since

$$\int_{-\infty}^{\infty} |g_2(x)|^p\,dx \le \int_{[x|\,|g(x)|<1]} |g(x)|^p\,dx < \infty,$$

and hence

$$\int_{-\infty}^{\infty} |g_2(x)|^2\,dx = \int_{-\infty}^{\infty} |g_2(x)|^{p+(2-p)}\,dx = \int_{[x|\,|g(x)|<1]} |g(x)|^{p+(2-p)}\,dx$$

$$< \int_{[x|\,|g(x)|<1]} |g(x)|^p\,dx < \infty.$$

The operator F so defined on $L_p(-\infty,\infty)$, $1 < p < 2$, may be called the Fourier transform as well.

§3. Convergence and summability

The problem of convergence of the "Fourier integral"

$$(3.1)\qquad S_R(f;x) \equiv \frac{1}{\sqrt{(2\pi)}}\int_{-R}^{R} F[f](\alpha)\, e^{-i\alpha x}d\alpha,\ R>0,\ f\in L_2(-\infty,\infty),$$

$$-\infty < x < \infty,$$

as $R\to\infty$, can be handled in the same way as in the L_1-case, (Ch.I, §§4, 7), once the basic formula (cf. (4.2), Ch.I)

$$(3.2)\qquad S_R(f;x) - f(x) = \frac{2}{\pi}\int_0^{\infty} g_x(t)\,\frac{\sin Rt}{t}\,dt,$$

$$g_x(t) = \tfrac{1}{2}[f(x+t) + f(x-t) - 2f(x)]$$

is established, which can be done by using Plancherel's theorem.

For let

$$F[\varphi](y) = \begin{cases} e^{-iyx}, & |y| < R,\\ 0, & |y| \geq R,\end{cases}\qquad R>0,\ -\infty<x<\infty.$$

Then we have

$$\varphi(u) = \frac{1}{\sqrt{(2\pi)}}\int_{-\infty}^{\infty} F[\varphi](y)\cdot e^{-iyu}dy = \frac{1}{\sqrt{(2\pi)}}\int_{-R}^{R} e^{-iy(x+u)}dy$$

$$= \left(\frac{2}{\pi}\right)^{1/2}\frac{\sin R(x+u)}{(x+u)},$$

and formula (2.25) gives

$$\int_{-R}^{R} F[f](\alpha)\cdot e^{-i\alpha x}d\alpha = \int_{-\infty}^{\infty} f(u)\varphi(-u)\,du = \int_{-\infty}^{\infty} f(u)\left(\frac{2}{\pi}\right)^{1/2}\frac{\sin R(x-u)}{(x-u)}\,du$$

$$= \left(\frac{2}{\pi}\right)^{1/2}\int_{-\infty}^{\infty} f(x-t)\,\frac{\sin Rt}{t}\,dt$$

$$= \left(\frac{2}{\pi}\right)^{1/2}\int_0^{\infty} [f(x+t) + f(x-t)]\,\frac{\sin Rt}{t}\,dt,$$

which leads to (3.2). The following analogue of Theorem 5 of Chapter I is a consequence, the proof being similar.

Theorem 2. If $f \in L_2(-\infty,\infty)$, *and* f *is of bounded variation in a neighbourhood of the point* $x \in (-\infty,\infty)$, *then*

$$\lim_{R\to\infty} S_R(f;x) \equiv \lim_{R\to\infty} \frac{1}{\sqrt{(2\pi)}} \int_{-R}^{R} F[f](\alpha)\cdot e^{-i\alpha x} d\alpha = \tfrac{1}{2}[f(x+0) - f(x-0)].$$

The problem of $(C,1)$ summability, for example, can again be handled without any new difficulty once the basic formula (cf. (7.2), Ch.I)

$$(3.3)\quad \frac{1}{\sqrt{(2\pi)}} \int_{-R}^{R} F[f](\alpha) \left(1 - \frac{|\alpha|}{R}\right) e^{-i\alpha x} d\alpha - f(x)$$

$$= \frac{1}{\pi} \int_{0}^{\infty} g_x(t)\, \frac{\sin^2(Rt/2)}{R(t/2)^2}\, dt$$

is established. Let

$$F[\varphi](y) = \begin{cases} \left(1 - \frac{|y|}{R}\right) e^{-ixy}, & |y| \leq R, \\ 0, & |y| > R. \end{cases}$$

Then we have

$$\varphi(u) = \frac{1}{\sqrt{(2\pi)}} \int_{-R}^{R} \left(1 - \frac{|y|}{R}\right) e^{-i(x+u)y} dy = \frac{4}{\sqrt{(2\pi)}}\, \frac{\sin^2[R(x+u)/2]}{R(x+u)^2}$$

(cf. Example 2, §1, Ch.I), so that formula (2.25) again gives

$$\frac{1}{\sqrt{(2\pi)}} \int_{-R}^{R} F[f](\alpha) \left(1 - \frac{|\alpha|}{R}\right) e^{-i\alpha x} d\alpha = \frac{2}{\pi} \int_{-\infty}^{\infty} f(u)\, \frac{\sin^2[R(x-u)/2]}{R(x-u)^2}\, du$$

$$= \frac{2}{\pi} \int_{-\infty}^{\infty} f(x-t)\, \frac{\sin^2(Rt/2)}{Rt^2}\, dt$$

$$= \frac{1}{2\pi} \int_{0}^{\infty} [f(x+t) + f(x-t)]\, \frac{\sin^2(Rt/2)}{R(t/2)^2}\, dt,$$

which leads to (3.3). Following the same lines of proof as in Theorem 10 of Chapter I, we obtain, for example,

Theorem 3. If $f \in L_2(-\infty,\infty)$, *then*

$$\lim_{R\to\infty} \frac{1}{\sqrt{(2\pi)}} \int_{-R}^{R} F[f](\alpha) \left(1 - \frac{|\alpha|}{R}\right) e^{-i\alpha x} d\alpha = f(x),$$

for almost all $x \in (-\infty,\infty)$.

We note that formula (2.24) plays the same rôle in the L_2-theory as the composition rule (cf. (1.13), Ch.I) did in the L_1-case. The problem of inversion in $L_2(-\infty,\infty)$ does not arise in the same form as in $L_1(-\infty,\infty)$, and the theorems on point-wise summability of the Fourier integral, while true, do not carry the same import.

To consider the question of summability in the L_2-norm, we need an extension of Theorem 2 of Chapter I, on the convolution of two functions in $L_1(-\infty,\infty)$.

(3.4) If $f \in L_1(-\infty,\infty)$, and $g \in L_p(-\infty,\infty)$, where $1 \le p \le 2$, then for almost every $x \in (-\infty,\infty)$, the function $f(x-y)\ g(y)$ belongs to $L_1(-\infty < y < \infty)$, and if we define $f*g$ by the relation

$$(f*g)(x) = \int_{-\infty}^{\infty} f(x-y)g(y)dy,$$

then

$$\|f*g\|_p \le \|f\|_1 \ \|g\|_p, \tag{3.5}$$

so that $f*g \in L_p(-\infty,\infty)$.

This is proved in the same way as Lemma 2.1 and Theorem 2, of Chapter I, except that one has to use Hölder's inequality instead of Schwarz's.

In particular, if $p = 2$, and $h \equiv f*g$, then $h \in L_2(-\infty,\infty)$, so that $F[h]$ is defined, as well as $F[f]$ (cf. (1.2), Ch.I), and $F[g]$. We shall now prove

<u>*Theorem 4.*</u> *If* $f \in L_1(-\infty,\infty)$, *and* $g \in L_2(-\infty,\infty)$, *and* $h = f*g$, *then* $h \in L_2(-\infty,\infty)$, *and*

$$F[h](x) = \sqrt{(2\pi)}\ F[f](x)\ F[g](x), \tag{3.6}$$

for almost all $x \in (-\infty,\infty)$.

<u>Proof</u>. Obviously $F[f]$ is bounded; in fact, $\sqrt{(2\pi)}\ |F[f](\alpha)| \le \|f\|_1 = M < \infty$, and if (g_n) is a sequence of functions belonging to Schwartz's space S, (cf. §3, Ch.I), such that $\|g-g_n\|_2 \to 0$, as $n \to \infty$, and if we set

$h_n = f*g_n$, then $h_n \in L_1(-\infty,\infty) \cap L_2(-\infty,\infty)$, and $F[h_n] = \sqrt{(2\pi)}\ F[f]\ F[g_n]$ by Theorem 2 of Chapter I. Further, because of (3.5), $\|h-h_n\|_2 = \|f*(g-g_n)\|_2 \to 0$, as $n \to \infty$. Since $\|F[h] - F[h_n]\|_2 = \|h-h_n\|_2 \to 0$, as $n \to \infty$, and

$$\|F[h_n] - \sqrt{(2\pi)}\ F[f]\ F[g]\|_2 = \|F[f]\ F[g_n] - F[f]\ F[g]\|_2\ (\sqrt{(2\pi)})$$

$$\le M\ \|F[g] - F[g_n]\|_2 = M\ \|g-g_n\|_2 \to 0,$$

as $n \to \infty$, we obtain (3.6).

The question of summability in the L_2-norm is answered by the following

Theorem 5. Let $f \in L_2(-\infty,\infty)$, $K \in L_2(-\infty,\infty)$, *and* $F[K] \equiv H \in L_1(-\infty,\infty) \cap L_2(-\infty,\infty)$, *H even. For* $R > 0$, *let* $H_R(y) = RH(Ry)$. *Then we have*

$$(3.7)\qquad \left\|\frac{1}{\sqrt{(2\pi)}}\ (f*H_R) - f\right\|_2 \to 0, \text{ as } R \to \infty.$$

Proof. By formula (2.25), we have

$$\int_{-\infty}^{\infty} F[f](y)\ K(\tfrac{y}{R})e^{-iyx}dy = \int_{-\infty}^{\infty} f(x+y)\ H_R(y)dy$$

$$= \int_{-\infty}^{\infty} f(x-y)H_R(y)dy = (f*H_R)(x).$$

By (3.5) we see that $f*H_R \in L_2(-\infty,\infty)$, for each $R > 0$. By using the properties of the L_2-modulus of continuity (see (1.17), Ch.I), we deduce as before (cf. Th.16, Ch.I) that

$$\left\|\frac{1}{\sqrt{(2\pi)}}\ (f*H_R) - f\right\|_2 \to 0, \text{ as } R \to \infty.$$

Remark. Clearly we can take for $K(\alpha)$ the Abel kernel $e^{-|\alpha|}$, and the Gauss kernel $e^{-\alpha^2}$, as we did in Chapter I.

§4. The closure of translations

Let $f \in L_2(-\infty,\infty)$, and let $\overline{S}_f$ denote the closure, in the L_2-norm, of the set S_f of all "translations" of the form $\sum_{k=1}^{m} c_k\ f(\cdot+t_k)$, where t_k

is real, and c_k complex (cf. §12, Ch.I). Then we have the following

Theorem 6 (Wiener). If $f \in L_2(-\infty,\infty)$, *with* $F[f](\alpha) \neq 0$ *for almost all* $\alpha \in (-\infty,\infty)$, *then* $\overline{S}_f = L_2(-\infty,\infty)$.

This can be obtained as a special case of another theorem which provides a sufficient condition for a function in $L_2(-\infty,\infty)$ to belong to $\overline{S}_f$.

Let $\mathbb{R}_1$ denote the set of all real numbers, and let $\varphi = F[f]$ where $f \in L_2(-\infty,\infty)$. Let E_φ denote a set in $\mathbb{R}_1$ with the property that, except for *null sets* (i.e. measurable sets of Lebesgue measure zero), $\varphi(\alpha) \neq 0$ for $\alpha \in E_\varphi$, and $\varphi(\alpha) = 0$ for $\alpha \in \mathbb{R}_1 - E_\varphi$. The set E_φ is, of course, measurable.

If $F(x) = \sum_{k=1}^{m} c_k f(x+t_k)$, F is not identically zero, and $\Phi = F[F]$, then

$$\Phi(\alpha) = \left\{c_1 e^{-i\alpha t_1} + \ldots + c_m e^{-i\alpha t_m}\right\}\varphi(\alpha),$$

and clearly $E_\Phi = E_\varphi$, except for null sets, and if $g \in \overline{S}_f$, with $F[g] = \psi$, then

(4.1) $$E_\psi < E_\varphi ,$$

which is defined to mean that $E_\psi = E_1 \cdot U \cdot E_2$, where $E_1 \subset E_\varphi$ and E_2 is a null set. But the converse is also true, as shown by the following

Theorem 7. If $f, g \in L_2(-\infty,\infty)$, *and* $F[f] = \varphi$, $F[g] = \psi$, *and if*

(4.2) $$E_\psi < E_\varphi ,$$

then $g \in \overline{S}_f$.

Proof. Clearly $\overline{S}_f$ is a closed, linear subspace of $L_2(-\infty,\infty)$, which is a separable Hilbert space. Given any $g \in L_2(-\infty,\infty)$, there exists an element $g^0 \in \overline{S}_f$, such that

(4.3) $$g = g^0 + f^0,$$

with

(4.4) $$(f^0,h) \equiv \int_{-\infty}^{\infty} f^0(x)\, \overline{h}(x)\,dx = 0,$$

for *every* $h \in \overline{S}_f$. If $\varphi^O = F[f^O]$, we have again

$$E_{\varphi^O} < E_\varphi , \tag{4.5}$$

because of (4.1) and (4.2). If we now choose $h(x) = f(x+t)$, for a fixed, real t, in (4.4), then we have, by (2.24) and (2.22),

$$\int_{-\infty}^{\infty} \varphi^O(\alpha)\ \overline{\varphi(\alpha)}\ e^{i\alpha t} d\alpha = 0,$$

where $\varphi^O \cdot \varphi \in L_1(-\infty,\infty)$. Hence, by Theorem 7 of Chapter I,

$$\varphi^O(\alpha)\ \overline{\varphi(\alpha)} = 0, \tag{4.6}$$

for almost all $\alpha \in (-\infty,\infty)$. But we have (except for sets of measure zero) $\varphi^O(\alpha) = 0$ for $\alpha \in \mathbb{R}_1 - E_\varphi$, by (4.5), while $\varphi(\alpha) \neq 0$ for $\alpha \in E_\varphi$. Hence $\varphi^O(\alpha) = 0$ for almost every α. By Plancherel's theorem, $f^O(x) = 0$, for almost every x, which gives, because of (4.3), $g(x) = g^O(x)$ for almost every x, and hence $g \in \overline{S}_f$.

Proof of Theorem 6. If $E_\varphi = \mathbb{R}_1$, then automatically we have $E_\psi < E_\varphi$, for every $g \in L_2(-\infty,\infty)$. Theorem 7 then implies that $g \in \overline{S}_f$, for every $g \in L_2(-\infty,\infty)$.

§5. Heisenberg's inequality

We shall now prove an L_2-analogue of what is referred to as Heisenberg's inequality, originally proved by Weyl under somewhat stronger assumptions.

Theorem 8. Let $f \in L_2(-\infty,\infty)$. *Then we have*

$$\int_{-\infty}^{\infty} x^2 |f(x)|^2 dx \cdot \int_{-\infty}^{\infty} \alpha^2 |F[f](\alpha)|^2 d\alpha \geq \frac{1}{4}\ \|f\|_2^4 , \tag{5.1}$$

and the equality takes place only in case $f(x) = c\ e^{-kx^2}$, $k > 0$, $c \in \mathbb{C}$.

Proof. We may assume that

$$\int_{-\infty}^{\infty} x^2|f(x)|^2dx < \infty, \quad \int_{-\infty}^{\infty} \alpha^2|F[f](\alpha)|^2d\alpha < \infty,$$

for otherwise (5.1) is trivially true since neither term can be zero.

Let $f^*(\alpha)$ denote the inverse transform of $(-i\alpha\ F[f](\alpha))$, that is to say

$$f^*(\alpha) = \check{F}[-i\alpha\ F[f](\alpha)]. \tag{5.2}$$

By Plancherel's theorem, $f^* \in L_2(-\infty,\infty)$. The left-hand side of (5.1) equals

$$\int_{-\infty}^{\infty} x^2|f(x)|^2dx \int_{-\infty}^{\infty} \alpha^2|F[f](\alpha)|^2d\alpha$$

$$= \int_{-\infty}^{\infty} x^2|f(x)|^2dx \int_{-\infty}^{\infty} |(-i\alpha\ F[f](\alpha))|^2d\alpha$$

$$= \int_{-\infty}^{\infty} x^2|f(x)|^2dx \int_{-\infty}^{\infty} |f^*(\alpha)|^2d\alpha,$$

by (5.2) and Plancherel's theorem. By Schwarz's inequality, we have

$$\int_{-\infty}^{\infty} x^2|f(x)|^2dx \int_{-\infty}^{\infty} |f^*(x)|^2dx \geq \left[\int_{-\infty}^{\infty} \frac{x(f^*\bar{f} + \overline{f^*}f)}{2}\,dx\right]^2, \tag{5.3}$$

since $\operatorname{Re}[x\ f^*(x)\bar{f}(x)] = \frac{1}{2}\,x(f^*\bar{f} + \overline{f^*}f)$.

To prove the theorem it suffices therefore to show that

$$\left[\int_{-\infty}^{\infty} x(f^*\bar{f} + \overline{f^*}f)\,dx\right]^2 = \|f\|_2^4. \tag{5.4}$$

Let $f_n \in S$ - Schwartz's space - be so chosen that

$$\lim_{n\to\infty} \int_{-\infty}^{\infty} (1+\alpha^2)\,|F[f_n](\alpha) - F[f](\alpha)|^2d\alpha = 0. \tag{5.5}$$

This is indeed possible, since S is a dense subset of $L_2(-\infty,\infty)$. Since $\alpha|F[f](\alpha)| \in L_2(-\infty,\infty)$ by assumption, and $F[f] \in L_2(-\infty,\infty)$, we have $(1+\alpha^2)^{1/2}F[f](\alpha) \in L_2(-\infty,\infty)$. There exists a sequence (g_n), $g_n \in S$, such that

$$\int_{-\infty}^{\infty} |g_n(\alpha) - (1+\alpha^2)^{1/2}F[f](\alpha)|^2d\alpha \to 0, \text{ as } n\to\infty,$$

or

$$\int_{-\infty}^{\infty} (1+\alpha^2) \left|\frac{g_n(\alpha)}{(1+\alpha^2)^{1/2}} - F[f](\alpha)\right|^2 d\alpha \to 0, \text{ as } n \to \infty.$$

Choose f_n such that

$$F[f_n](\alpha) = \frac{g_n(\alpha)}{(1+\alpha^2)^{1/2}} ,$$

noting that $F[f_n] \in S$ implies that $f_n \in S$ (cf. Ch.I, (3.7)).

Now by Plancherel's theorem, and (5.5),

$$\|f_n-f\|_2^2 + \|f_n'-f^*\|_2^2 = \int_{-\infty}^{\infty} (1+\alpha^2)|F[f_n](\alpha) - F[f](\alpha)|^2 d\alpha \to 0,$$

as $n \to \infty$,

where f_n' denotes the derivative of f_n, so that $F[f_n'](\alpha) = -i\alpha \cdot F[f_n](\alpha)$, and f^* is defined as in (5.2). It follows that

(5.6) $$\|f_n-f\|_2 \to 0, \quad \|f_n'-f^*\|_2 \to 0, \text{ as } n \to \infty.$$

Now

$$\int_{-\infty}^{\infty} |F[f_n](\alpha) - F[f](\alpha)|d\alpha$$

$$= \int_{-\infty}^{\infty} \frac{(1+\alpha^2)^{1/2}}{(1+\alpha^2)^{1/2}} |F[f_n](\alpha) - F[f](\alpha)|d\alpha$$

$$\le \left[\int_{-\infty}^{\infty} \frac{d\alpha}{(1+\alpha^2)}\right]^{1/2} \left[\int_{-\infty}^{\infty} (1+\alpha^2)|F[f_n](\alpha) - F[f](\alpha)|^2 d\alpha\right]^{1/2}$$

$$= B_n, \text{ say,}$$

where $B_n \to 0$, as $n \to \infty$. Hence

(5.7) $$F[f_n] - F[f] \in L_1(-\infty,\infty), \text{ for } n = 1,2,\ldots .$$

But $f_n - f \in L_2(-\infty,\infty)$, so that

$$f_n(x)-f(x) = \lim_{\substack{R\to\infty \\ (L_2\text{-norm})}} \frac{1}{\sqrt{(2\pi)}} \int_{-R}^{R} [F[f_n](\alpha) - F[f](\alpha)]e^{-i\alpha x}d\alpha$$

$$= \frac{1}{\sqrt{(2\pi)}} \int_{-\infty}^{\infty} [F[f_n](\alpha) - F[f](\alpha)]e^{-i\alpha x}d\alpha,$$

almost everywhere, by (5.7). Hence we have, almost everywhere,

$$(5.7)' \qquad |f_n(x)-f(x)| \leq ||F[f_n] - F[f]||_1 \leq B_n \to 0, \text{ as } n\to\infty.$$

Therefore there exists $c > 0$, independent of x, such that

$$(5.8) \qquad |f_n(x)-f(x)| < c < \infty, \text{ almost everywhere.}$$

Since $f^* \in L_2(-\infty,\infty)$ by definition, we have $f^* \in L_1(-R,R)$ for $0 < R < \infty$. And

$$\left|\int_{-R}^{R} (f_n'\overline{f}_n - f^*\overline{f})(x)dx\right|$$

$$\leq \int_{-R}^{R} |f_n'\overline{f}_n - f_n'\overline{f} + f_n'\overline{f} - f^*\overline{f}|dx$$

$$\leq \int_{-R}^{R} |f_n'|\cdot|\overline{f}_n-\overline{f}|dx + \int_{-R}^{R} |\overline{f}|\cdot|f_n'-f^*|dx$$

$$(5.9) \qquad \leq B_n \int_{-R}^{R} |f_n'|dx + \left(\int_{-R}^{R} |f|^2dx\right)^{1/2} \left(\int_{-R}^{R} |f_n'-f^*|^2dx\right)^{1/2}.$$

Since $f \in L_2(-\infty,\infty)$, and $||f_n'-f^*||_2 \to 0$, as $n\to\infty$, by (5.6), the second term on the right-hand side of (5.9) tends to zero as $n\to\infty$. The first term on the right-hand side is

$$(5.10) \qquad B_n \int_{-R}^{R} |f_n'-f^*+f^*|dx \leq B_n \int_{-R}^{R} |f_n'-f^*|dx + B_n \int_{-R}^{R} |f^*|\, dx$$

$$\to 0, \text{ as } n\to\infty, \text{ (for fixed } R > 0),$$

since $||f_n'-f^*||_2 \to 0$. From (5.9) and (5.10) we have

$$(5.11) \qquad \lim_{n\to\infty} \int_{-R}^{R} f_n'\overline{f}_n\, dx = \int_{-R}^{R} f^*\overline{f}\, dx,$$

and similarly also

$$\int_{-R}^{R} x\, f_n'(x)\overline{f}_n(x)dx \to \int_{-R}^{R} x\, f^*(x)\overline{f}(x)dx, \text{ as } n\to\infty,$$

and

$$\int_{-R}^{R} x\, \overline{f_n'}(x)f_n(x)dx \to \int_{-R}^{R} x\, \overline{f^*}(x)f(x)dx, \text{ as } n\to\infty.$$

Hence

$$\lim_{R\to\infty}\lim_{n\to\infty}\int_{-R}^{R} x(f_n'\overline{f}_n + \overline{f_n'}f_n)dx = \int_{-\infty}^{\infty} x(f^*\overline{f} + \overline{f^*}f)dx, \tag{5.12}$$

where

$$\left|\int_{-\infty}^{\infty} x(f^*\overline{f} + \overline{f^*}f)dx\right| < \infty, \tag{5.12}'$$

by (5.3) together with the hypotheses: $x|f(x)|\in L_2(-\infty < x < \infty)$, $\alpha|F[f](\alpha)| \in L_2(-\infty < \alpha < \infty)$.

Now

$$\int_{-R}^{R} x(f_n'(x)\overline{f}_n(x) + \overline{f_n'}(x)f_n(x))dx = \int_{-R}^{R} x(|f_n(x)|^2)'dx$$

the dash denoting the derivative, and

$$\lim_{R\to\infty}\lim_{n\to\infty}\int_{-R}^{R} x(|f_n(x)|^2)'dx$$

$$= \lim_{R\to\infty}\lim_{n\to\infty}\left\{\Big[x|f_n(x)|^2\Big]_{-R}^{R} - \int_{-R}^{R}|f_n(x)|^2dx\right\}$$

$$= \lim_{R\to\infty}\left\{R[|f(R)|^2 + |f(-R)|^2]\right\} - \|f\|_2^2 , \tag{5.13}$$

since $f_n\to f$ almost everywhere, as $n\to\infty$, by (5.7)', and $\|f_n - f\|_2\to 0$, as $n\to\infty$ [assuming, as we may, that $R\to\infty$ through values which are not in the exceptional null set in (5.7)']. The left-hand side of (5.13) is finite by (5.12) and (5.12)'; so is $\|f\|_2$; hence

$$\lim_{R\to\infty} R\left\{|f(R)|^2 + |f(-R)|^2\right\}$$

is finite, and ≥ 0. If the limit is $\delta > 0$, then there exists R_0 such that we have

$$\int_{R_0}^{\infty} \left\{|f(R)|^2 + |f(-R)|^2\right\} dR > \frac{\delta}{2} \int_{R_0}^{\infty} \frac{dR}{R} = \infty,$$

which contradicts the assumption: $f \in L_2(-\infty,\infty)$. Hence

$$\lim_{R\to\infty} R\left\{|f(R)|^2 + |f(-R)|^2\right\} = 0,$$

and this, taken together with (5.13) and (5.12), yields

$$\left[\int_{-\infty}^{\infty} x(f^*(x)\overline{f}(x) + \overline{f^*}(x)f(x))dx\right]^2 = \|f\|_2^4,$$

so that (5.4) is proved, hence the *in*equality in the theorem.

In order to determine when the inequality becomes an equality, we note (by the first application of Schwarz's inequality just before (5.3)) that

$$\int_{-\infty}^{\infty} x^2|f(x)|^2 dx \int_{-\infty}^{\infty} \alpha^2 |F[f](\alpha)|^2 d\alpha = \left[\int_{-\infty}^{\infty} |x\ f(x)f^*(x)|^2 dx\right]^2$$

if (and only if) $f^*(x) = K\ x\ \overline{f}(x)$ almost everywhere, K being a complex constant. Here f^* is defined as in (5.2). In fact, f^* is the derivative *almost everywhere* of f. For by (5.6), we have

$$\|f_n' - f^*\|_2 \to 0, \text{ as } n\to\infty,$$

which implies that

$$\lim_{n\to\infty} \int_0^x f_n'(y)dy = \int_0^x f^*(y)dy$$

over any finite interval $[0,x]$. But the left-hand side equals $\lim_{n\to\infty} [f_n(x) - f_n(0)] = f(x) - f(0)$, almost everywhere, by (5.7)'.
Hence f equals, almost everywhere, an absolutely continuous function, and

$$f^*(x) = \frac{d}{dx}[f(x)], \text{ for almost all } x \in (-\infty,\infty). \tag{5.14}$$

With this identification, (5.1) becomes an equality if

$$\frac{d}{dx}(f(x)) = K\ x\ \overline{f}(x), \quad (K \text{ a complex constant})$$

or $\frac{1}{x}\frac{d}{dx}(f(x)) = K\,\overline{f}(x)$, or $\frac{1}{x}\frac{d}{dx}\left(\frac{1}{x}\frac{d}{dx}f(x)\right) = \frac{1}{x}\frac{d}{dx}(K\,\overline{f}) = |K|^2 f(x)$, that is to say, $\left(\frac{1}{x}\frac{d}{dx}\right)^2 f(x) - |K|^2 f(x) = 0$, or $(D^2 - |K|^2)f = 0$, where $D = \left(\frac{1}{x}\frac{d}{dx}\right)$, which implies that $(D - |K|)(D + |K|)f = 0$.

If $(D + |K|)f = 0$, then $\frac{1}{x}\frac{df}{dx} = -|K|f$, or $\frac{f'}{f} = -|K|x$, or $\log f = -|K|\frac{x^2}{2} + c$, or

$$f(x) = c_1\, e^{-|K|x^2/2} \in L_2(-\infty,\infty). \tag{5.15}$$

If $(D - |K|)f = 0$, then $f(x) = c_2\, e^{+|K|x^2/2} \notin L_2(-\infty,\infty)$.

The function $f(x) = e^{-|K|x^2/2}$ actually satisfies the *equality* in (5.1), for if $f^*(x) = K\,x\overline{f}(x)$, then we have

$$\int_{-\infty}^{\infty} x^2|f(x)|^2dx \int_{-\infty}^{\infty} \alpha^2|F[f](\alpha)|^2d\alpha = \left(\int_{-\infty}^{\infty} |xf^*(x)\overline{f}(x)|dx\right)^2$$

$$= \left(\int_{-\infty}^{\infty} |K|x^2|f(x)|^2dx\right)^2 = |K|^2\left(\int_{-\infty}^{\infty} x^2|f(x)|^2dx\right)^2,$$

and if $f(x) = e^{-|K|x^2/2}$, the last expression is

$$|K|^2\left(\int_{-\infty}^{\infty} x^2|f(x)|^2dx\right)^2 = |K|^2\left(\int_{-\infty}^{\infty} x^2\, e^{-|K|x^2}dx\right)^2$$

$$= |K|^2\left(\int_{-\infty}^{\infty} x\,\frac{-2|K|x}{-2|K|}\,e^{-|K|x^2}dx\right)^2$$

$$= |K|^2\left(\int_{-\infty}^{\infty} \frac{e^{-|K|x^2}}{2|K|}\,dx\right)^2 \quad \text{(by partial integration)}$$

$$= |K|^2\left(\frac{1}{2|K|^{3/2}}\int_{-\infty}^{\infty} e^{-t^2}dt\right)^2$$

$$= |K|^2\,\frac{1}{4|K|^3}\,(\sqrt{\pi})^2 = \frac{\pi}{4|K|},$$

while

$$\frac{1}{4}\,\|f\|_2^4 = \frac{1}{4}\left(\int_{-\infty}^{\infty}|f(x)|^2dx\right)^2 = \frac{1}{4}\left(\int_{-\infty}^{\infty}e^{-|K|x^2}dx\right)^2$$

$$= \frac{1}{4}\left(\frac{1}{\sqrt{(|K|)}}\int_{-\infty}^{\infty} e^{-t^2}dt\right)^2$$

$$= \frac{1}{4|K|}(\sqrt{\pi})^2 = \frac{\pi}{4|K|} .$$

§6. Hardy's theorem

We have noted that the function $f(x) = e^{-x^2}$ has the special property that

$$\int_{-\infty}^{\infty} e^{-x^2} e^{i\alpha x}dx = \sqrt{\pi}\, e^{-\alpha^2/4}, \quad -\infty < \alpha < \infty,$$

(§1, Ch.I), which implies that

$$\frac{1}{\sqrt{(2\pi)}}\int_{-\infty}^{\infty} e^{-x^2/2} e^{i\alpha x}dx = e^{-\alpha^2/2}.$$

Hardy has shown that this property of the function $e^{-x^2/2}$, together with its order of magnitude, characterize the function in the sense of the following

Theorem 9. Let $f(x)$ *be a measurable function defined on* $-\infty < x < \infty$. *Let* $f(x) = O(e^{-x^2/2})$, *as* $|x| \to \infty$, *and*

$$F[f](\alpha) = \frac{1}{\sqrt{(2\pi)}}\int_{-\infty}^{\infty} f(x)e^{i\alpha x}dx = O\left(e^{-\alpha^2/2}\right),$$

as $|\alpha| \to \infty$. *Then*

$$f(x) = c\, e^{-x^2/2},$$

where c *is a complex number.*

For the proof we need to apply the principle of Phragmén-Lindelöf, which may be viewed as an extension of the maximum principle.

(6.1) *Theorem of Phragmén-Lindelöf. Let* $f(z)$ *be a non-constant holomorphic function of the complex variable* z $(= re^{i\theta})$, *in the domain* D *defined by the relations*

$$D: r \geq 0, \ -\frac{\pi}{2\alpha} \leq \theta \leq \frac{\pi}{2\alpha}, \ \alpha > \frac{1}{2}$$

and let

$$|f(z)| \leq M < \infty, \text{ for } r \geq 0, \ \theta = \pm \frac{\pi}{2\alpha},$$

and

$$|f(z)| < K\, e^{r^\beta}, \ \beta < \alpha, \ z \in \overline{D},$$

where $\overline{D}$ *denotes the closure of* D, *and the constant* K *is independent of* z. *Then we have*

$$|f(z)| < M, \ z \in D.$$

Proof. Consider the function

$$F(z) = e^{-\varepsilon z^\gamma} f(z), \ \beta < \gamma < \alpha, \ \varepsilon > 0.$$

We have

$$|F(z)| = e^{-\varepsilon r^\gamma \cos\gamma\theta} |f(z)|.$$

On the lines $\theta = \pm \frac{\pi}{2\alpha}$, we have $\cos\gamma\theta > 0$, since $\gamma < \alpha$. Hence on these lines

$$|F(z)| \leq |f(z)| \leq M.$$

Further on the arc defined by $|z| = R > 0$, $|\theta| \leq \frac{\pi}{2\alpha}$, we have

$$|F(z)| \leq e^{-\varepsilon R^\gamma \cos(\frac{1}{2}\gamma\frac{\pi}{\alpha})} |f(z)|$$

$$< K \cdot e^{R^\beta - \varepsilon R^\gamma \cos(\frac{1}{2}\frac{\gamma\pi}{\alpha})} \to 0, \text{ as } R \to \infty,$$

since $\beta < \gamma < \alpha$. Hence for $R \geq R_0 > 0$, we have $|F(z)| \leq M$, on that arc. By the maximum principle, we have

$$|F(z)| \leq M, \text{ for } 0 \leq r \leq R, \ |\theta| \leq \frac{\pi}{2\alpha}.$$

It follows that

$$|f(z)| \leq M \cdot e^{\varepsilon r^\gamma}, \ z \in D, \ \varepsilon > 0.$$

On letting $\varepsilon \downarrow 0$, we get the required result.

(6.2) *Corollary. Take* $\alpha = 2$, $\beta = 1$.

Proof of Theorem 9. Let $z = x + iy$, where x and y are real, $i = \sqrt{-1}$. Consider the function defined by

$$(6.3)\qquad \tilde{f}(z) = \frac{1}{\sqrt{(2\pi)}} \int_{-\infty}^{\infty} f(t)e^{izt}dt.$$

The integral converges absolutely and uniformly in every strip $-\infty < -A \le \operatorname{Im} z \le A < \infty$, since

$$\frac{1}{\sqrt{(2\pi)}} \int_{-\infty}^{\infty} |f(t)e^{izt}|dt \le K_1 \int_{-\infty}^{\infty} e^{-\frac{1}{2}t^2-yt} dt = K_1 \int_{-\infty}^{\infty} e^{-\frac{1}{2}(t+y)^2+y^2/2} dt$$

$$(6.4)\qquad = K_2\, e^{\frac{1}{2}y^2}, \text{ say.}$$

Hence $\tilde{f}(z)$ is an *entire* function of z.

Since

$$f(x) = \tfrac{1}{2}(f(x) + f(-x)) + \tfrac{1}{2}(f(x) - f(-x)) = f_1(x) + f_2(x),$$

say, where f_1 is an *even* function, and f_2 is odd, and f_1, f_2 satisfy the same conditions as f, we shall consider the case f even, and f odd, separately.

Let f be even. Then $\tilde{f}$ is also even, so that

$$\tilde{f}(z) = \sum_{n=0}^{\infty} a_n z^{2n},$$

and

$$(6.5)\qquad \varphi(z) := \tilde{f}(\sqrt{z}) = \sum_{n=0}^{\infty} a_n z^n$$

is an entire function.

Set $z = r\, e^{i\theta}$, so that $x = r \cos\theta$, $y = r \sin\theta$, $\sqrt{z} = \sqrt{r}\, e^{i\theta/2}$, and $\operatorname{Im}(\sqrt{z}) = r^{1/2}\sin(\theta/2)$.

From (6.4) we have

$$(6.6)\qquad |\varphi(z)| = |\tilde{f}(\sqrt{z})| \le K_2\, e^{\frac{1}{2}r \sin^2(\theta/2)} \le K_2\, e^{\frac{1}{2}r},$$

for all z.

For z real, and positive, say $z = r > 0$, we have

$$|\varphi(r)| = |\tilde{f}(\sqrt{r})| \le K_3 e^{-\frac{1}{2}r}, \tag{6.7}$$

by hypothesis. Choose $M > \max(K_2, K_3)$, and α such that $0 < \alpha < \pi$, and define the function

$$\omega(z,\alpha) \equiv \omega(r,\theta,\alpha) = \exp\left[\frac{iz}{2} \cdot \frac{e^{-i\alpha/2}}{\sin(\alpha/2)}\right]$$

$$= \exp\left[\frac{ir}{2} \cdot \frac{e^{i(\theta-\alpha/2)}}{\sin(\alpha/2)}\right]$$

$$= \exp\left[\frac{ir}{2\sin(\alpha/2)}\left\{\cos(\theta - \tfrac{\alpha}{2}) + i\sin(\theta - \tfrac{\alpha}{2})\right\}\right]$$

$$= \exp\left[-\frac{r}{2} \cdot \frac{\sin(\theta-\alpha/2)}{\sin(\alpha/2)} + i\,\frac{r\cos(\theta-\alpha/2)}{2\sin(\alpha/2)}\right].$$

Now

$$|\omega(r,0,\alpha)| = e^{\frac{1}{2}r}, \quad |\omega(r,\alpha,\alpha)| = e^{-\frac{1}{2}r}.$$

Hence (on writing $\varphi(r,\theta) \equiv \varphi(z)$, $\varphi(r) = \varphi(r,0)$), by (6.7), we have

$$|\omega(r,0,\alpha)\cdot\varphi(r,0)| \le M,$$

and

$$|\omega(r,\alpha,\alpha)\cdot\varphi(r,\alpha)| \le M, \quad \text{by (6.6)}.$$

By the Phragmén-Lindelöf principle, we obtain

$$|\omega(z)\ \varphi(z)| \le M, \text{ for } 0 \le \theta \le \alpha,\ 0 < \alpha < \pi;$$

that is to say

$$|\varphi(z)| \le M \exp\left[\frac{r}{2} \cdot \frac{\sin(\theta-\alpha/2)}{\sin(\alpha/2)}\right].$$

If we keep θ fixed, and let $\alpha \uparrow \pi$, then

$$|\varphi(z)| \le M\cdot e^{-\frac{r}{2}\cos\theta}, \text{ for } 0 \le \theta < \pi.$$

By continuity this holds also for $\theta = \pi$.

Similarly we consider the half-plane $-\pi < \theta < 0$, and obtain

$$|e^{\frac{1}{2}z} \varphi(z)| \leq M, \quad \text{for all (finite) } z.$$

Since φ is entire, $e^{\frac{1}{2}z} \varphi(z)$ reduces to a constant, therefore

$\varphi(z) = c_1 \cdot e^{-\frac{1}{2}z}$, or $\tilde{f}(z) = c \cdot e^{-\frac{1}{2}z^2}$, and the theorem follows.

If f is *odd*, then $\tilde{f}$ is *odd*, hence $\tilde{f}(0) = 0$, so that $\frac{\tilde{f}(z)}{z}$ is an even, entire function. By what we have just proved, $\tilde{f}(z) = c_2\, z\, e^{-\frac{1}{2}z^2}$. But for $z = x$ real, $\tilde{f}(x) = O\left(e^{-\frac{1}{2}x^2}\right)$, by hypothesis. This is possible only if $c_2 = 0$ or $\tilde{f}(z) = f(z) \equiv 0$.

§7. The theorem of Paley and Wiener

A fundamental theorem due to Paley and Wiener enables us to give a characterization of the Fourier transforms of functions belonging to $L_2(-\infty,\infty)$ which vanish outside a finite interval, in terms of entire functions of exponential type in the complex plane.

An entire function f(z) of the complex variable z is said to be *of exponential type*, if

$$f(z) = O\left(e^{A|z|}\right), \quad \text{as } z \to \infty, \tag{7.1}$$

for some (finite) $A > 0$. The lower bound of such numbers A is called the *type* of f; it is non-negative.

We denote by E^σ, $\sigma \geq 0$, the class of entire functions *of type at most* σ. Thus if $f \in E^\sigma$, then

$$f(z) = O\left(e^{(\sigma+\varepsilon)|z|}\right), \quad \text{as } z \to \infty, \tag{7.2}$$

for *every* $\varepsilon > 0$.

<u>*Theorem 10*</u> *(Paley and Wiener). Let* $0 < A < \infty$. *Then we have*

(7.3) $$F(x) = \frac{1}{\sqrt{(2\pi)}} \int_{-A}^{A} f(u)e^{ixu}du, \quad -\infty < x < \infty,$$

for some $f \in L_2(-A,A)$, *if and only if* $F(x) \in L_2(-\infty < x < \infty)$ *and* F *can be extended to the complex plane as a function of the class* E^A.

<u>Proof</u> (First part). If (7.3) holds with $f(u) \in L_2(-A < u < A)$, then $f(u) \in L_1(-A < u < A)$, and for complex z,

$$F(z) = \frac{1}{\sqrt{(2\pi)}} \int_{-A}^{A} f(u)e^{izu}du \qquad (z = x+iy)$$

is an entire function of z, with

$$|F(z)| \le \frac{1}{\sqrt{(2\pi)}} e^{A|z|} \int_{-A}^{A} |f(u)|du = O(e^{A|z|}),$$

so that $F \in E^A$. Further the Fourier transform of f, where $f(u)$ is defined to be zero for $|u| > A$, is F, so that by Plancherel's theorem, $||F||_2 = ||f||_2 < \infty$.

(Second part). Let $F(x) \in L_2(-\infty < x < \infty)$, and for complex z, let $F(z) \in E^A$. Then by Plancherel's theorem,

(7.4) $$f(x) = \lim_{\substack{R\to\infty \\ (L_2\text{-norm})}} \frac{1}{\sqrt{(2\pi)}} \int_{-R}^{R} F(u)e^{-ixu}du$$

belongs to $L_2(-\infty < x < \infty)$. We shall prove that $f(x)$ vanishes almost everywhere for $|x| > A > 0$, so that $f(x) \in L_2(-A < x < A)$.

For complex z, let

(7.5) $$g(z) = \int_{-\frac{1}{2}}^{\frac{1}{2}} F(u-z)du.$$

Then

(7.6) $g(z)$ is an entire function of z,

such that, for $\varepsilon > 0$,

$$|g(z)| \le \int_{-\frac{1}{2}}^{\frac{1}{2}} |F(u-z)|du = O\left(\int_{-\frac{1}{2}}^{\frac{1}{2}} e^{(A+\varepsilon)(|z|+|u|)}du\right)$$

$$(7.7) \qquad = O\left(e^{(A+\varepsilon)|z|} \int_{-\frac{1}{2}}^{\frac{1}{2}} e^{(A+\varepsilon)|u|}du\right)$$

$$= O\left(e^{(A+\varepsilon)|z|}\right),$$

since $F \in E^A$. Further, for real x, we have

$$(7.8) \qquad |g(x)|^2 \le \int_{-\frac{1}{2}}^{\frac{1}{2}} |F(u-x)|^2 du \le \|F\|_2^2 < \infty,$$

and

$$\int_{-\infty}^{\infty} |g(x)|^2 dx \le \int_{-\infty}^{\infty} \left\{\int_{-\frac{1}{2}}^{\frac{1}{2}} |F(u-x)|^2 du\right\}dx = \int_{-\frac{1}{2}}^{\frac{1}{2}} du \int_{-\infty}^{\infty} |F(u-x)|^2 dx$$

$$(7.9) \qquad = \int_{-\frac{1}{2}}^{\frac{1}{2}} \|F\|_2^2 du = \|F\|_2^2 < \infty,$$

so that g(x) is *bounded, with* $g(x) \in L_2(-\infty < x < \infty)$.

Next let

$$(7.10) \qquad G(z) = e^{iBz} g(z), \quad z \text{ complex}, \ B > A > 0.$$

Then G(z) is an entire function of *exponential type*, by (7.6) and (7.7). Further if z is real, and z = x, then

$$(7.11) \qquad |G(x)| = |g(x)| = O(1), \quad \text{by (7.8)},$$

and if z = ib, b > 0, then by (7.7), we have

$$|G(ib)| = |e^{-Bb} g(ib)| = O\left(e^{-Bb+(A+\varepsilon)b}\right) = O\left(e^{-b(B-A-\varepsilon)}\right) \to 0, \quad \text{as } b \to +\infty,$$

if ε is chosen sufficiently small, since B > A. In particular,

(7.12) $|G(ib)| = O(1)$, as $b \to +\infty$, for $B > A$.

Because of (7.11) and (7.12) and the Phragmén-Lindelöf principle (6.1), it follows that $|e^{iBz}g(z)| = O(1)$, for $z = Re^{i\theta}$, $0 \le \theta \le \frac{\pi}{2}$, or $|g(Re^{i\theta})| < c_1 e^{BR\sin\theta}$, $0 \le \theta \le \frac{\pi}{2}$. Similarly $|g(Re^{i\theta})| < c_2 e^{BR\sin\theta}$, for $\frac{\pi}{2} < \theta \le \pi$. Hence we have

(7.13) $$|g(Re^{i\theta})| = O(e^{BR\sin\theta}),\ R > 0,\ 0 \le \theta \le \pi.$$

Let $L > 0$, and $x > B > A > 0$, and consider the integral

$$\int_{-R}^{R} \frac{e^{ixu}}{1-Liu}\, g(u)du,$$

where R is large enough to ensure that $LR > 1$.

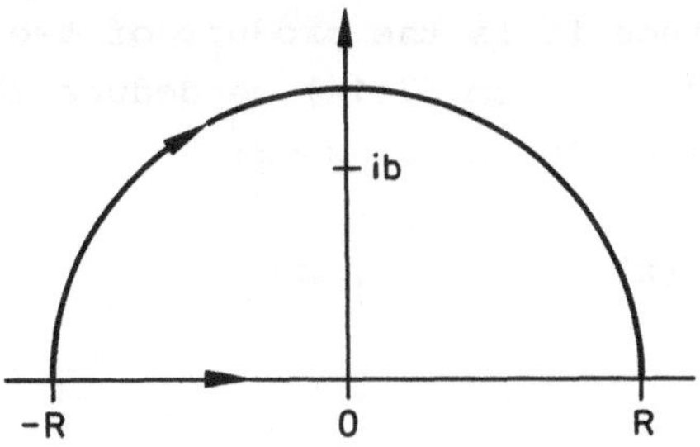

By Cauchy's theorem applied to the semicircular contour defined by $|\mathrm{Re}\, z| \le R$, $\mathrm{Im}\, z = 0$; and $\mathrm{Im}\, z > 0$, $|z| = R$; (see Fig.), we have

$$\int_{-R}^{R} \frac{e^{ixu}}{1-Liu}\, g(u)du = -i \int_{0}^{\pi} \frac{e^{ixRe^{i\theta}}}{1-LiRe^{i\theta}}\, g(Re^{i\theta})\ Re^{i\theta}d\theta$$

so that

(7.14) $$\left|\int_{-R}^{R} \frac{e^{ixu}g(u)}{1-Liu}\, du\right| \le \frac{c\cdot R}{LR-1} \int_{0}^{\pi} e^{-xR\sin\theta + BR\sin\theta}d\theta \to 0,\ \text{as } R \to \infty,$$

since $LR > 1$, $x > B$, by choice, $\sin\theta \ge 0$ for $0 \le \theta \le \pi$, and

(7.15) $$\int_{0}^{\pi} e^{-aR\sin\theta}d\theta = O(\tfrac{1}{R}),\ \text{for any } a > 0,$$

which can be seen as follows. We have

$$\int_0^\pi e^{-aR\sin\theta}d\theta = \int_0^{\pi/2} + \int_{\pi/2}^{\pi} = I_1 + I_2, \text{ say,}$$

where

$$|I_1| \le \left|\int_0^{\pi/2} e^{-cR\theta}d\theta\right| = \left|\frac{-e^{-cR\pi/2}}{cR} + \frac{1}{cR}\right| = O(\tfrac{1}{R}), \; c = \frac{2a}{\pi} > 0,$$

since $\frac{2}{\pi} \le \frac{\sin\theta}{\theta} \le 1$, for $0 \le \theta \le \frac{\pi}{2}$, so that $-aR\sin\theta \le \frac{-2a}{\pi}R\theta = -cR\theta$. And

$$\int_{\pi/2}^{\pi} e^{-aR\sin\theta}d\theta = -\int_{\pi/2}^{0} e^{-aR\sin\varphi}d\varphi = I_1, \text{ (with } \theta = \pi-\varphi, \; \sin\theta = \sin\varphi).$$

Let

(7.16) $$g_{-L}(u) = \frac{g(u)}{1-Liu}, \; L>0, \; -\infty<u<\infty.$$

Then $g_{-L} \in L_1(-\infty,\infty)$, since it is the product of two functions each of which belongs to $L_2(-\infty,\infty)$. From (7.14) we deduce that the Fourier transform of g_{-L} vanishes for $L>0$, $x>B$;

(7.17) $$F[g_{-L}](x) = 0, \; L>0, \; x>B.$$

We have, however,

(7.18) $$\int_{-\infty}^{\infty} \left|\frac{g(u)}{1-Liu} - g(u)\right|^2 du \to 0, \text{ as } L \downarrow 0,$$

by Lebesgue's theorem on dominated convergence (cf. (7.8)). Hence, by Plancherel's theorem,

(7.19) $$\int_{-\infty}^{\infty} |F[g_{-L}](u) - F[g](u)|^2du \to 0, \text{ as } L \downarrow 0.$$

From (7.17) it follows that

(7.20) $$F[g](x) = 0, \text{ almost everywhere, for } x>B.$$

We have, however, for real v,

(7.21) $$g(v) = \int_{-1/2}^{1/2} F(u-v)du = \frac{1}{\sqrt{(2\pi)}} \int_{-\infty}^{\infty} f(x)\left(\frac{\sin x/2}{x/2}\right) e^{-ivx}dx,$$

by (2.26). (Note that $\check{F}[f](x) = F[f](-x)$). Since

$f(x)\,\frac{\sin x/2}{x/2} \in L_1(-\infty,\infty)\cdot\cap\cdot L_2(-\infty,\infty)$, we deduce from (7.21) that

$$F[g](x) = f(x)\,\frac{\sin x/2}{x/2},$$

for almost all $x \in (-\infty,\infty)$. Since (7.20) holds for *every* $B > A$, we conclude that $f(x) = 0$ *for almost every* $x > A$.

Similarly we show that $f(x) = 0$ for almost every $x < -A$. For we have only to consider $G(-ib)$, $b > 0$, instead of $G(ib)$, and note that, after (7.7),

$$(7.22)\qquad |G(-ib)| = |e^{Bb}g(-ib)| = O\left(e^{b(A+\varepsilon+B)}\right),\ b > 0,$$

for every $\varepsilon > 0$, and if $B < -A < 0$, then for ε sufficiently small, we have

$$(7.23)\qquad |G(-ib)| = O(1),\ \text{as } b \to +\infty,\ B < -A < 0,$$

corresponding to (7.12). Again by the Phragmén-Lindelöf principle applied to a semi-circular domain in the lower half-plane, we obtain

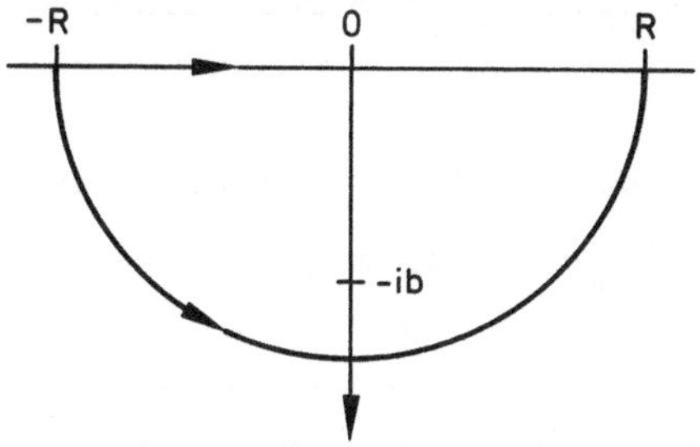

$$(7.24)\qquad |g(Re^{i\theta})| = O(e^{BR\sin\theta}),\ -\pi \le \theta \le 0,\ R > 0,$$

corresponding to (7.13). And we have

$$\left|\int_{-R}^{R} \frac{e^{ixu}g(u)}{1+Liu}\,du\right| \le \left|i\int_{-\pi}^{0} \frac{e^{ixRe^{i\theta}}g(Re^{i\theta})\,Re^{i\theta}}{1+LiRe^{i\theta}}\,d\theta\right|$$

$$\le \frac{cR}{LR-1}\int_{-\pi}^{0} e^{-(x-B)R\sin\theta}d\theta = \frac{cR}{LR-1}\int_{0}^{\pi} e^{(x-B)R\sin\theta}d\theta,$$

which tends to zero as $R \to \infty$ if $x < B$, c being a constant. This corresponds to (7.14). And we deduce (as in (7.17)) that

(7.25) $\qquad F[g_L](x) = 0, \text{ for } L > 0,\ x < B < -A < 0,$

where $g_L(x) = \frac{g(x)}{1+Lix} \in L_1(-\infty < x < \infty)$. It then follows, as before, that $f(x) = 0$ for almost every $x < -A$, and the second part of the theorem is proved.

Combining the Paley-Wiener theorem with the Riemann-Lebesgue theorem (Ch.I, Th.1), we obtain the following

(7.26) <u>*Corollary.*</u> *If* $F(z)$ *is an entire function of exponential type, with* $z = x+iy$, *and* $F(x) \in L_2(-\infty < x < \infty)$, *then* $F(x) \to 0$ *as* $|x| \to \infty$.

§8. Fourier series in $L_2(a,b)$

Let $\varphi_0(x)$, $\varphi_1(x)$, $\varphi_2(x)$,..., be complex-valued, non-null functions (that is, functions which are not almost everywhere equal to zero) defined on the interval (a,b) of the real line. We say that $\{\varphi_n\}$ is an *orthogonal* set if

$$\int_a^b \varphi_m(x)\overline{\varphi}_n(x)\,dx = \begin{cases} 0, & \text{for } m \neq n;\ m,n = 0,1,2,\ldots, \\ \lambda_m > 0, & \text{for } m = n. \end{cases}$$

We note that $\varphi_m(x) \in L_2(a < x < b)$, and that no $\varphi_m(x)$ can vanish identically, since that would imply that $\lambda_m = 0$.

If, in addition, $\lambda_m = 1$ for $m = 0,1,2,\ldots$, we call $\{\varphi_n\}$ an *orthonormal* set.

If $\{\varphi_n\}$ is orthogonal, then clearly $\left\{\frac{\varphi_n}{\lambda_n^{1/2}}\right\}$ is orthonormal.

Given $f \in L_2(a,b)$, let

$$c_n = \frac{1}{\lambda_n}\int_a^b f(x)\overline{\varphi}_n(x)\,dx,$$

for any integer $n \geq 0$, We call c_n the nth *Fourier coefficient* of f, and write

(8.1) $$f \sim (c_n).$$

We call the series

$$\sum_{n=0}^{\infty} c_n \varphi_n(x)$$

the *Fourier (orthogonal) series* of f, relative to (φ_n), and indicate that relation by writing

(8.2) $$f(x) \sim \sum_{n=0}^{\infty} c_n \varphi_n(x) .$$

The set $\{e^{inx}\}$, $n = 0, \pm 1, \pm 2, \ldots$, is *orthogonal* on $(0,2\pi)$, while the set $\{e^{inx} (2\pi)^{-1/2}\}$ is *orthonormal*, since

$$\int_{\alpha}^{\alpha+2\pi} e^{imx} \cdot e^{-inx} \, dx = \begin{cases} 0, & m \neq n, \\ 2\pi, & m = n, \end{cases}$$

for any real α.

In order to have the *equality*

$$f(x) = \sum_{n=0}^{\infty} c_n \varphi_n(x), \text{ for } any \ f \in L_2(a,b),$$

it is necessary that the set $\{\varphi_n\}$ should be "complete", in the sense that if we add a new, non-null function ψ to the set, then the resulting set is *no longer* orthogonal. Otherwise there would exist a non-null function, namely ψ itself, with *all* its Fourier coefficients equal to zero.

We thus *define* a set (φ_n), $\varphi_n \in L_2(a,b)$, to be *complete*, if there exists no non-null function in $L_2(a,b)$ which is orthogonal to φ_n for every $n \geq 0$. In other words,

$$\int_a^b f(x) \, \overline{\varphi}_n(x) dx = 0, \quad n = 0,1,2,\ldots,$$

implies that $f(x) = 0$, for almost every $x \in (a,b)$.

It follows that if $\{\varphi_n\}$ is *complete, and orthonormal,* and the functions $f,g \in L_2(a,b)$ have the *same* Fourier coefficients, then f is *equivalent* to g (cf.§1, Ch.I), that is to say that f equals g almost everywhere in (a,b). In this sense, the Fourier series of f is unique.

Let $f \in L_2(a,b)$, with $||f||_2 = \left(\int_a^b |f(x)|^2 dx\right)^{1/2}$. Let (φ_n) be an *orthonormal* set in $L_2(a,b)$, and let c_n denote the nth Fourier coefficient of f relative to (φ_n). By a *polynomial* in the φ_n, of rank k, where k is a non-negative integer, is meant an expression of the form

$$\Phi_k(x) = \sum_{m=0}^{k} \gamma_m \varphi_m(x), \tag{8.3}$$

where all γ's are complex numbers. We call

$$f_k(x) = \sum_{m=0}^{k} c_m \varphi_m(x) \tag{8.4}$$

the *Fourier polynomial*, of rank k, *of the function* f.

With this notation we have the following identity:

$$||f-\Phi_n||_2^2 = ||f||_2^2 - \sum_{m=0}^{n} |c_m|^2 + \sum_{m=0}^{n} |c_m-\gamma_m|^2, \tag{8.5}$$

and, in particular,

$$||f-f_n||_2^2 = ||f||_2^2 - \sum_{m=0}^{n} |c_m|^2. \tag{8.6}$$

To prove (8.5) we note that

$$\int_a^b f(x)\,\overline{\Phi}_n(x)dx = \sum_{m=0}^{n} c_m \overline{\gamma}_m, \quad \int_a^b |\Phi_n(x)|^2 dx = \sum_{m=0}^{n} |\gamma_m|^2,$$

and

$$||f-\Phi_n||_2^2 = \int_a^b \{f(x) - \Phi_n(x)\}\{\overline{f}(x) - \overline{\Phi}_n(x)\}dx$$

$$= ||f||_2^2 - \sum_{m=0}^{n} c_m \overline{\gamma}_m - \sum_{m=0}^{n} \overline{c}_m \gamma_m + \sum_{m=0}^{n} |\gamma_m|^2$$

$$= ||f||_2^2 - \sum_{m=0}^{n} |c_m|^2 + \sum_{m=0}^{n} (c_m-\gamma_m)(\overline{c}_m-\overline{\gamma}_m). \tag{8.7}$$

By (8.6), $\sum_{m=0}^{n} |c_m|^2 \le ||f||_2^2$, and on letting $n \to \infty$, we obtain *Bessel's inequality*, namely

(8.8) $$\sum_{n=0}^{\infty} |c_n|^2 \leq ||f||_2^2 = \int_a^b |f(x)|^2 dx.$$

From (8.5) we conclude that

(8.9) *the best approximation, in the* L_2*-norm, to* $f \in L_2(a,b)$ *by polynomials* Φ_n*, of a given rank* n*, is provided by the Fourier polynomial* f_n *of* f.

An important result on the Fourier series of functions in $L_2(a,b)$ is the following

(8.10) *The Riesz-Fischer theorem. Given any sequence* (c_n) *of complex numbers, such that* $\sum_{n=0}^{\infty} |c_n|^2 < \infty$*, there exists a function* $f \in L_2(a,b)$*, such that the* c_n*'s are the Fourier coefficients of* f*, relative to the given orthonormal set* $(\varphi_n)_{n=0}^{\infty}$*. We have further:*

(8.11) $$\int_a^b |f_n(x) - f(x)|^2 dx \to 0 \quad \textit{as } n \to \infty,$$

and

(8.12) $$\int_a^b |f(x)|^2 dx = \sum_{n=0}^{\infty} |c_n|^2.$$

Proof. Let $f_m \equiv f_m(x) = \sum_{k=0}^{m} c_k \varphi_k(x) \equiv \sum_{k=0}^{m} c_k \varphi_k$. If $n > m$, then

$$\int_a^b |f_m(x) - f_n(x)|^2 dx = \sum_{k=m+1}^{n} |c_k|^2 \to 0, \text{ as } m,n \to \infty,$$

by the hypothesis on the c_n. Since the function space $L_2(a,b)$ is complete (cf. §2) (which is not to be confused with the set (φ_n) being complete, which is *not* assumed here), there exists a function $f \in L_2(a,b)$, such that

(8.13) $$\int_a^b |f_n(x) - f(x)|^2 dx \to 0, \text{ as } n \to \infty.$$

It follows that

$$\int_a^b f(x) \overline{\varphi}_m(x) dx = \lim_{n\to\infty} \int_a^b f_n(x) \overline{\varphi}_m(x) dx = \lim_{n\to\infty} c_m = c_m,$$

so that c_m *is* the mth Fourier coefficient of f, relative to (φ_n).

(Note that if $\|f-f_n\|_2 \to 0$, and $\|g-g_n\|_2 \to 0$, then $\|fg - f_n g_n\|_1 \to 0$.) Hence, by (8.5), we have

$$\|f-f_n\|_2^2 = \|f\|_2^2 - \sum_{m=0}^{n} |c_m|^2,$$

and (8.13) now implies (8.12).

Remark. We note that by (8.6), (8.11) is equivalent to (8.12). We next make use of the completeness of (φ_n).

(8.14) *Parseval's theorem. (A) If* $f \in L_2(a,b)$, *and* (φ_n) *is a complete, orthonormal set in* $L_2(a,b)$, *and* c_n *denotes the* n^{th} *Fourier coefficient of* f, *relative to the* (φ_n), *then*

(8.11)' $$\int_a^b |f_n(x) - f(x)|^2 dx \to 0, \text{ as } n \to \infty,$$

and

(8.12)' $$\int_a^b |f(x)|^2 dx = \sum_{n=0}^{\infty} |c_n|^2.$$

(B). If $f,g \in L_2(a,b)$, $f \sim (c_n)$ *and* $g \sim (d_n)$, *(as defined in (8.1)), then*

$$\int_a^b f(x)\overline{g}(x)dx = \sum_{n=0}^{\infty} c_n \overline{d}_n ,$$

the series on the right converging absolutely.

Proof. (A) By Bessel's inequality we have $\sum_{n=0}^{\infty} |c_n|^2 < \infty$, and the Riesz-Fischer theorem gives an $f \in L_2(a,b)$ satisfying (8.11) and (8.12). The completeness of (φ_n) implies that that f differs from the given f only on a set of measure zero in (a,b), and (8.11)' and (8.12)' follow.

(B) We have

$$\int_a^b f_n(x)\overline{g}(x)dx = \sum_{m=0}^{n} c_m \int_a^b \varphi_m(x)\overline{g}(x)dx = \sum_{m=0}^{n} c_m \overline{d}_m.$$

Since $\|f-f_n\|_2 \to 0$, as $n \to \infty$, by (8.11)' it follows that

$$\int_a^b f_n(x)\overline{g}(x)dx \to \int_a^b f(x)\overline{g}(x)dx, \quad \text{as } n \to \infty,$$

and since $\sum_{n=0}^{\infty} |c_n|^2$, and $\sum_{n=0}^{\infty} |d_n|^2$, converge by (8.12)', we have $\sum_{m=0}^{\infty} |c_m \bar{d}_m| < \infty$.

The closure and completeness of (φ_n) in $L_2(a,b)$

(8.15) A set of functions (φ_n), $\varphi_n \in L_2(a,b)$, is said to be *closed* in $L_2(a,b)$, if the polynomials in φ_n (see (8.3)) form a *dense* subset of $L_2(a,b)$ in the L_2-norm.

(8.16) The orthonormal set (φ_n) is closed in $L_2(a,b)$, if and only if it is complete in $L_2(a,b)$.

Proof. If (φ_n) is complete, then given any $f \in L_2(a,b)$, there exists, by (8.11)', a sequence (f_n) of Fourier polynomials of f, such that $\|f-f_n\|_2 \to 0$, as $n \to \infty$. Hence the set (φ_n) is closed.

Conversely let the set (φ_n) be closed in $L_2(a,b)$, and $f \in L_2(a,b)$, $f \sim (c_n)$, and $c_n = 0$ for all $n \geq 0$. Then there exists a sequence of polynomials in the φ_n, say Φ_n, such that $\|\Phi_n - f\|_2 \to 0$, as $n \to \infty$. By (8.9), we have $\|f_n - f\|_2 \to 0$, as $n \to \infty$, where f_n is the Fourier polynomial of f of rank n. But $f_n = 0$, since all the c_n's are zero, hence $\|f\|_2 = 0$, or f is zero almost everywhere in (a,b). It follows that (φ_n) is complete in $L_2(a,b)$.

Remarks

1. Let $\varphi_n(x) = \frac{1}{\sqrt{(2\pi)}} e^{inx}$, $|x| \leq \pi$, $n = 0, \pm 1, \pm 2, \ldots,$.

   It is a classical result that (φ_n) is complete in $L_1(-\pi,\pi)$, hence also in $L_2(-\pi,\pi)$.

2. Let
$$\varphi_0(x) = \begin{cases} +1, & \text{for } 0 < x < \frac{1}{2} \\ -1, & \text{for } \frac{1}{2} < x < 1, \end{cases}$$
   and $\varphi_0(0) = \varphi_0(\frac{1}{2}) = 0$, with $\varphi(x) = \varphi(x+1)$. Let $\varphi_n(x) = \varphi_0(2^n x)$ for $n = 0, 1, 2, \ldots$. Then (φ_n) is orthonormal over (0,1).

   The function $\varphi_n(x)$ takes alternately the values +1, and −1 in the intervals $(0,2^{-n-1})$, $(2^{-n-1}, 2\cdot 2^{-n-1})$, $(2\cdot 2^{-n-1}, 3\cdot 2^{-n-1})$, ... If

$m > n$, the integral of $\varphi_m \varphi_n$ over any of these intervals is zero, and the set (φ_n) is obviously orthonormal. It is not complete, since the function $\psi(x) = 1$ may be added to it. (Note that the set $\{\psi\} \cup \{\varphi_n\}_{n\geq 0}$ is not complete.) If we define

$$\text{sign}\ \alpha = 1 \text{ for } \alpha > 0, \ \text{sign}\ \alpha = -1 \text{ for } \alpha < 0, \text{ and}$$
$$\text{sign}\ \alpha = 0 \text{ for } \alpha = 0,$$

then we have $\varphi_n(x) = \text{sign} \sin 2^{n+1}\pi x$. The functions (φ_n) are known as *Rademacher's functions*.

3. The condition of orthonormality in (8.16) is *not* necessary.

§9. Hardy's interpolation formula

As an application of the Paley-Wiener theorem one can obtain some special formulae of interpolation at integer points for entire functions of exponential type, which help to determine the value of the function at an arbitrary point in the complex plane in terms of its values at the integer points on the real axis. The notions concerning Fourier series outlined in the previous section enter into the proofs.

If $F(z)$ is an entire function of z belonging to the class E^σ, $\sigma > 0$, as defined in (7.2), then $F(\frac{\pi z}{\sigma}) \in E^\pi$, so that the study of functions in E^σ, $\sigma > 0$, may be reduced to that of functions in E^π.

Theorem 11 (Hardy). Let $F(z) \in E^\pi$, $z = x+iy$, *and*

$$\int_{-\infty}^{\infty} |F(x)|^2 dx < \infty. \tag{9.1}$$

Then we have

$$F(z) = \frac{\sin \pi z}{\pi} \sum_{n=-\infty}^{\infty} (-1)^n \frac{F(n)}{z-n}. \tag{9.2}$$

Proof. If we define

$$\varphi_n(x) = \begin{cases} \frac{1}{\sqrt{(2\pi)}} e^{inx}, & |x| \leq \pi, \\ 0, & |x| > \pi, \end{cases}$$

for $n = 0, \pm 1, \pm 2, \ldots$, then the set (φ_n) is orthonormal over $(-\infty,\infty)$, but not complete on $(-\infty,\infty)$ (since any function which vanishes on $(-\pi,\pi)$ has all its Fourier coefficients equal to zero). It follows that the set $\{\check{F}[\varphi_n]\}$ is also orthonormal (by (2.24)). Now

$$\check{F}[\varphi_n](\alpha) = \frac{1}{\sqrt{(2\pi)}} \int_{-\infty}^{\infty} \varphi_n(x) e^{-i\alpha x} dx = \frac{1}{2\pi} \int_{-\pi}^{\pi} e^{i(n-\alpha)x} dx$$

$$(9.3) \qquad = \frac{\sin(\alpha-n)\pi}{\pi(\alpha-n)} = (-1)^n \frac{\sin(\alpha\pi)}{\pi(\alpha-n)} .$$

Since $F(x) \in L_2(-\infty,\infty)$, let

$$(9.4) \qquad F(x) \sim \sum_{n=-\infty}^{\infty} a_n \check{F}[\varphi_n](x) = \sum_{n=-\infty}^{\infty} a_n (-1)^n \frac{\sin(\pi x)}{\pi(x-n)} ,$$

where

$$a_n = \frac{1}{\sqrt{(2\pi)}} \int_{-\infty}^{\infty} F(x) \overline{\check{F}[\varphi_n](x)} dx$$

is the nth Fourier coefficient of F relative to $(\check{F}[\varphi_n])$. [Since $(\check{F}[\varphi_n])$ is not complete, the series in (9.4) does not, in general, represent F].

By the Paley-Wiener theorem, $F[F](x) = 0$ for $|x| > \pi$, hence $F[F] \in L_1(-\infty,\infty) \cdot \cap \cdot L_2(-\infty,\infty)$. (Note that $\check{F}[F](\alpha) = F[F](-\alpha)$). Since (φ_n) is complete in $L_2(-\pi,\pi)$, given $\varepsilon > 0$, there exists a polynomial in the φ_n's, with complex coefficients α_n, such that

$$\| F[F] - \sum_{-N}^{N} \alpha_n \varphi_n \|_2 < \varepsilon,$$

since φ_n vanishes outside $(-\pi,\pi)$ for every $n \geq 0$. By Plancherel's theorem we have

$$\| F - \sum_{-N}^{N} \alpha_n \check{F}[\varphi_n] \|_2 < \varepsilon.$$

This is equivalent to saying that

$$(9.5) \qquad \|F\|_2^2 = \sum_{n=-\infty}^{\infty} |a_n|^2 < \infty.$$

(See (8.16) and the Remark after the Riesz-Fischer theorem).

The series

$$\sum_{n=-\infty}^{\infty} \frac{\{\sin(\pi x)\}^2}{(x-n)^2}$$

is uniformly convergent on $(-\infty,\infty)$. By Schwarz's inequality, and (9.5), it follows that the series

$$\sum_{n=-\infty}^{\infty} a_n(-1)^n \frac{\sin \pi x}{\pi(x-n)}$$

converges uniformly on $(-\infty,\infty)$. Since $F \in E^\pi$ by hypothesis, F is continuous, hence

$$(9.6) \qquad F(x) = \sum_{n=-\infty}^{\infty} a_n(-1)^n \frac{\sin \pi x}{\pi(x-n)} = \frac{\sin \pi x}{\pi} \sum_{n=-\infty}^{\infty} \frac{a_n(-1)^n}{x-n},$$

for $-\infty < x < \infty$.

If we put $x = n$, we get

$$a_n = F(n), \; n = 0, \pm 1, \pm 2, \ldots .$$

The series on the right-hand side of (9.6) converges absolutely and uniformly also with a complex z in place of the real x, provided that $-\infty < -A \leq \operatorname{Im} z \leq A < +\infty$, hence represents an entire function of z, which coincides with F(z) because of (9.6), and (9.2) follows.

Theorem 11 can be used to prove an interpolation formula where the assumption that $F(x) \in L_2(-\infty < x < \infty)$ is replaced by the boundedness of F(x).

Theorem 12. If $F(z) \in E^\pi$, $z = x+iy$, *and* F(x) *is bounded for* $-\infty < x < \infty$, *then we have*

$$(9.7) \qquad F(z) = \frac{\sin \pi z}{\pi}\left\{F'(0) + \frac{F(0)}{z} + \sum_{\substack{n=-\infty \\ n\neq 0}}^{\infty} (-1)^n F(n)\left(\frac{1}{z-n} + \frac{1}{n}\right)\right\},$$

where F' *denotes the derivative of* F.

Proof. If we set $G(z) = \left(\frac{F(z) - F(0)}{z}\right)$, then $G(z) \in E^\pi$, and $G(x) \in L_2(-\infty < x < \infty)$. Theorem 11 then gives

$$G(z) = \frac{\sin \pi z}{\pi} \sum_{n=-\infty}^{\infty} \frac{(-1)^n G(n)}{z-n},$$

so that

(9.8) $$F(z) - F(0) = \frac{z\sin\pi z}{\pi} \sum_{\substack{n=-\infty \\ n\neq 0}}^{\infty} \frac{(-1)^n (F(n)-F(0))}{n(z-n)} + F'(0)\frac{\sin\pi z}{\pi},$$

since $G(0) = F'(0)$. We have, however,

$$\frac{\pi}{\sin\pi z} = \frac{1}{z} + \sum_{\substack{n=-\infty \\ n\neq 0}}^{\infty} (-1)^n \left\{\frac{1}{z-n} + \frac{1}{n}\right\} = \frac{1}{z} + z \sum_{\substack{n=-\infty \\ n\neq 0}}^{\infty} \frac{(-1)^n}{n(z-n)},$$

so that

$$F(0) = F(0)\frac{\sin\pi z}{\pi}\frac{\pi}{\sin\pi z} = F(0)\frac{\sin\pi z}{\pi}\left(\frac{1}{z} + z \sum_{\substack{n=-\infty \\ n\neq 0}}^{\infty} \frac{(-1)^n}{n(z-n)}\right),$$

and if we use this in (9.8) we obtain (9.7).

§10. Two inequalities due to S. Bernstein

The interpolation formula obtained in Theorem 12 can be used to prove two important inequalities originally proved by S. Bernstein.

Theorem 13 *(S. Bernstein). If* $F(z) \in E^{\sigma}$, $\sigma > 0$, $z = x+iy$, *and* $F(x)$ *is bounded for* $-\infty < x < \infty$, *with* $M = \sup |F(x)|$, *then we have*

(10.1) $$|F'(x)| \leq \sigma M, \quad -\infty < x < \infty,$$

the dash denoting the derivative. Equality occurs in (10.1) if and only if

(10.2) $$F(z) = a\,e^{i\sigma z} + b\,e^{-i\sigma z},$$

where a *and* b *are complex numbers.*

Proof. It is sufficient to prove the theorem for $\sigma = \pi$, for we can otherwise study $F(\frac{z\pi}{\sigma})$.

By taking $z = x$ in (9.7), and differentiating once, we get

(10.3) $$F'(x) = \cos\pi x\, F_1(x) + \frac{\sin\pi x}{\pi} \sum_{n=-\infty}^{\infty} \frac{(-1)^{n-1} F(n)}{(x-n)^2},$$

where

$$F_1(x) = F'(0) + \frac{F(0)}{x} + \sum_{\substack{n=-\infty \\ n\neq 0}}^{\infty} (-1)^n F(n)\left\{\frac{1}{x-n} + \frac{1}{n}\right\},$$

the differentiation being justified by the fact that the differentiated series converges uniformly. On setting $x = \frac{1}{2}$, we get

$$(10.4) \qquad F'(\tfrac{1}{2}) = \frac{4}{\pi} \sum_{n=-\infty}^{\infty} \frac{(-1)^{n-1} F(n)}{(2n-1)^2},$$

so that

$$(10.5) \qquad |F'(\tfrac{1}{2})| \leq \frac{4}{\pi} M \sum_{n=-\infty}^{\infty} \frac{1}{(2n-1)^2} = \frac{4M}{\pi}\frac{\pi^2}{4} = M\pi.$$

For *any* real x_0, consider

$$(10.6) \qquad G(z) = F(x_0 + z - \tfrac{1}{2}).$$

We have: $G \in E^{\pi}$, $|G(x)| \leq M$, $G'(\frac{1}{2}) = F'(x_0)$, $G(n) = F(x_0 + n - \frac{1}{2})$, and (10.5) implies that

$$|F'(x_0)| = |G'(\tfrac{1}{2})| \leq \pi M,$$

which proves the first part of the theorem.

To prove the second part, we apply formula (10.4) to the function G in (10.6), and obtain

$$G'(\tfrac{1}{2}) = F'(x_0) = \frac{4}{\pi} \sum_{n=-\infty}^{\infty} \frac{(-1)^{n-1} G(n)}{(2n-1)^2} = \frac{4}{\pi} \sum_{n=-\infty}^{\infty} \frac{(-1)^{n-1} F(x_0+n-\frac{1}{2})}{(2n-1)^2}.$$

On replacing x_0 by x, and n by $n+1$, we get

$$(10.7) \qquad F'(x) = \frac{4}{\pi} \sum_{n=-\infty}^{\infty} \frac{(-1)^n F(x+n+\frac{1}{2})}{(2n+1)^2}.$$

Let us suppose that for $x = x_1$ we have the equality

$$|F'(x_1)| = \pi M, \text{ or } F'(x_1) = \pi M e^{i\alpha}, \ \alpha \text{ real}.$$

Formula (10.7) then gives

$$(10.8)\qquad \frac{4}{\pi}\sum_{n=-\infty}^{\infty}\frac{\{(-1)^n F(x_1+n+\frac{1}{2}) - M\, e^{i\alpha}\}}{(2n+1)^2} = 0,$$

the series on the left-hand side being absolutely convergent, since F is bounded on the real axis.

Since $|F(x)| \le M$, we have $\mathrm{Re}\left[M-(-1)^n e^{-i\alpha}F(x_1+n+\frac{1}{2})\right] \ge 0$, and *if* $(-1)^n F(x_1+n+\frac{1}{2}) \neq M\, e^{+i\alpha}$, then $\mathrm{Re}\left[M-(-1)^n e^{-i\alpha}F(x_1+n+\frac{1}{2})\right] > 0$, hence (10.8) implies that

$$(-1)^n F(x_1+n+\tfrac{1}{2}) = M\, e^{i\alpha}, \text{ for } n = 0, \pm 1, \pm 2, \ldots\ .$$

If we set $H(z) = F(x_1+z+\frac{1}{2})$, then $H(0) = F(x_1+\frac{1}{2}) = M\, e^{i\alpha}$, and $H(n) = F(x_1+n+\frac{1}{2})$, so that

$$(10.9)\qquad (-1)^n H(n) = H(0).$$

If we apply Theorem 12 to the function H(z), we get

$$H(z) = \frac{\sin\pi z}{\pi}\left\{H'(0) + \frac{H(0)}{z} + \sum_{\substack{n=-\infty\\ n\neq 0}}^{\infty} H(0)\left(\frac{1}{z-n}+\frac{1}{n}\right)\right\},$$

because of (10.9). We have, however,

$$\pi\cot\pi z = \frac{1}{z} + \sum_{\substack{n=-\infty\\ n\neq 0}}^{\infty}\left(\frac{1}{z-n}+\frac{1}{n}\right).$$

Hence

$$\begin{aligned} H(z) &= \frac{\sin\pi z}{\pi}\left[H'(0) + H(0)\cdot\pi\cot\pi z\right] \\ &= A\cos\pi z + B\sin\pi z, \text{ say,} \\ &= A_1 e^{\pi z i} + B_1 e^{-\pi z i}. \end{aligned}$$

Since $F(z) = H(z-x_1-\frac{1}{2})$, it is proved that if equality occurs in (10.1), then (10.2) holds.

We have to show that if (10.2) holds, then equality occurs in (10.1) for some x, $-\infty < x < \infty$.

If $F(x) = a\, e^{i\sigma x} + b\, e^{-i\sigma x}$, and $a = |a|e^{i\alpha}$, $b = |b|e^{i\beta}$, $\sigma > 0$, then

$F(x) = e^{i(\alpha+\sigma x)}(|a| + |b|\, e^{i(\beta-\alpha)-2i\sigma x})$, so that for

$x = x_1 = \frac{(\beta-\alpha)-\pi}{2\sigma}$, we have $|F'(x_1)| = \sigma(|a| + |b|) = \sigma \operatorname{Max}_x |F(x)|$.

Remarks. By letting $\sigma \downarrow 0$ in (10.1), we deduce that the only functions $F \in E^0$, which are bounded on the real axis, are constants.

On the other hand, it follows from a theorem of Siegel that if $F \in E^0$, and $F(x) \in L_1(-\infty < x < \infty)$, then $F \equiv 0$.

Theorem 14 (S. Bernstein). Let $T(x) = \sum_{k=-n}^{n} c_k\, e^{ikx}$, *where* x *is real and* c_k *complex. Then the inequality*

$$|T(x)| \le M$$

implies that

$$|T'(x)| \le nM,$$

and this inequality becomes an equality if and only if

$$T(x) = a\, e^{inx} + b\, e^{-inx},$$

where a *and* b *are complex numbers.*

Proof. We have only to note that for complex z, T(z) is an entire function, with $T \in E^n$, while T(x) is bounded, and apply Theorem 13.

§11. Several variables

The proof of Plancherel's theorem in several variables is not essentially different from that in one variable. If E_k denotes the real Euclidean space of k dimensions, let S denote the Schwartz space of infinitely differentiable functions on E_k, such that for any $\alpha = (\alpha_1,\dots,\alpha_k)$, $\beta = (\beta_1,\dots,\beta_k)$, where $\alpha_1,\alpha_2,\dots,\alpha_k$ and $\beta_1,\beta_2,\dots,\beta_k$ are non-negative integers,

$$\sup_{x\in E_k} |x^\alpha (D^\beta f)(x)| < \infty,$$

where $x^{\alpha} = x_1^{\alpha_1} \ldots x_k^{\alpha_k}$, and $D^{\beta}f = \left(\frac{\partial}{\partial x_1}\right)^{\beta_1} \ldots \left(\frac{\partial}{\partial x_k}\right)^{\beta_k} f$. Then S is a dense subspace of $L_p(E_k)$, $1 \leq p < \infty$. For $f \in S$ we define the Fourier transform $F[f]$ by the relation

$$F[f](x) = (2\pi)^{-k/2} \int_{E_k} f(t)\, e^{i<x,t>} dt, \quad x \in E_k.$$

Then $F[f] \in S$, and $f \to F[f]$ is a one-to-one mapping of S onto S (as in the case of one variable, cf. §2). If $f, g \in S$, then $f*g \in S$. We further have

$$\int_{E_k} f(x)\, F[g](x)dx = \int_{E_k} F[f](y)\, g(y)dy,$$

and

$$(f,g) = (F[f], F[g]).$$

If $f \in L_2(E_k)$, there exists a sequence (f_n) of functions belonging to S, such that $||f-f_n||_2 \to 0$, as $n \to \infty$, and $||F[f_n] - F[f_m]||_2 = ||f_m-f_n||_2$, which implies that $||F[f_m] - F[f_n]||_2 \to 0$, as $m,n \to \infty$. Since $L_2(E_k)$ is complete, there exists $g \in L_2(E_k)$, such that $||g - F[f_m]||_2 \to 0$, as $m \to \infty$. And $||g||_2 = \lim_{m\to\infty} ||F[f_m]||_2 = \lim_{m\to\infty} ||f_m||_2 = ||f||_2$. We define g to be the *Fourier transform* of $f \in L_2(E_k)$. It is independent of the approximating sequence, and is defined almost everywhere. We denote it by $F[f]$.

We similarly define

$$\check{F}[f](x) = (2\pi)^{-k/2} \int_{E_k} f(t) e^{-i<x,t>} dt, \quad \text{for } f \in S,$$

and extend the definition to all of $L_2(E_k)$. It follows, as in (2.11), that

$$\check{F}\Big[F[f]\Big] = f = F\Big[\check{F}[f]\Big],$$

so that $f \to F[f]$ is a linear mapping of $L_2(E_k)$ onto itself; it is also isometric. The proof that the definitions of the Fourier transform on $L_2(E_k)$, and on $L_1(E_k)$ coincide on $L_1(E_k) \cap L_2(E_k)$ follows as in the case of E_1.

Chapter III. Fourier-Stieltjes transforms (one variable)

§1. Basic properties

We assume as known the fundamentals of the theory of Riemann-Stieltjes integrals.

Let $F(y)$ be a function of bounded variation for $-\infty < y < \infty$. For x real, let

$$(1.1) \qquad \varphi(x) = \int_{-\infty}^{\infty} e^{ixy} dF(y) \equiv \lim_{R\to\infty} \int_{-R}^{R} e^{ixy} dF(y), \quad (R > 0).$$

We call φ the *Fourier-Stieltjes transform* of F, or the Fourier transform of dF, and denote it sometimes by the symbol $\widehat{dF}$.

If, in particular,

$$(1.2) \qquad F(y) = \int_{-\infty}^{y} f(t)dt, \quad f \in L_1(-\infty,\infty),$$

then (1.1) reduces to the Fourier transform on $L_1(-\infty,\infty)$ studied in Chapter I.

The integral in (1.1) converges absolutely and uniformly and $\varphi(x)$ is a bounded, continuous function of x defined for every x in $(-\infty,\infty)$. We have only to note that

$$|\varphi(x)| \leq \int_{-\infty}^{\infty} |dF(y)| < \infty,$$

and for any real $h \neq 0$,

$$|\varphi(x+h) - \varphi(x)| \leq \int_{-\infty}^{\infty} \left|e^{ihy}-1\right| \cdot |dF(y)|$$

$$\leq |h| \int_{|y|<R} |y| \cdot |dF(y)| + 2 \int_{|y|\geq R} |dF(y)| = I_1 + I_2,$$

say, where $R > 0$. Given any $\varepsilon > 0$, one can choose R so large that $|I_2| < \varepsilon/2$, and h so small that $|I_1| < \varepsilon/2$.

But unlike the Fourier transform on $L_1(-\infty,\infty)$, $\varphi(x)$ does *not* necessarily tend to zero as $|x| \to \infty$. For example, if $F(x) = 1$, for $x > 0$; $F(x) = -1$, for $x < 0$; and $F(0) = 0$, then $\varphi(x) = 2$.

<u>*Theorem 1*</u>. *Let* $F(x)$ *be of bounded variation in* $(-\infty,\infty)$, *with*

$$F(x) = \frac{1}{2}\{F(x+0) + F(x-0)\}, \quad \textit{for all } x, \tag{1.3}$$

and

$$\varphi(x) = \int_{-\infty}^{\infty} e^{ixy}\, dF(y). \tag{1.4}$$

Then we have

$$F(x) - F(0) = \frac{1}{2\pi} \lim_{R\to\infty} \int_{-R}^{R} \varphi(t)\, \frac{e^{-itx}-1}{-it}\, dt \tag{1.5}$$

$$\equiv \frac{1}{2\pi} \int_{-\infty}^{\infty} \varphi(t)\, \frac{e^{-itx}-1}{-it}\, dt,$$

so that φ *determines* F *up to an arbitrary, additive constant.*

<u>Proof</u>. If h is real, and fixed, and $h \neq 0$, then (1.4) gives

$$\varphi(x)\, e^{-ihx} = \int_{-\infty}^{\infty} e^{ix(y-h)}\, dF(y) = \int_{-\infty}^{\infty} e^{ixy}\, dF(y+h). \tag{1.6}$$

From this and (1.4) we get

$$\varphi(x)\left[e^{-ihx}-1\right] = \int_{-\infty}^{\infty} e^{ixy}\, dG(y), \quad G(y) = F(y+h) - F(y). \tag{1.7}$$

Here G is of bounded variation in $(-\infty,\infty)$, and $G \in L_1(-\infty,\infty)$. For if we suppose $h \geq 0$, h fixed; and suppose that F is *non-decreasing* (since F is expressible as the difference of two non-decreasing, bounded functions, if it is real-valued; and if it is complex-valued, one can consider the real and imaginary parts separately), then $G(y) \geq 0$, and we have for $R > 0$,

$$\int_{-R}^{R} G(x)\,dx = \int_{-R}^{R} \{F(x+h) - F(x)\}dx = \left(\int_{R}^{R+h} - \int_{-R}^{-R+h}\right)F(x)\,dx$$

$$= \int_0^h F(x+R)dx - \int_{-h}^0 F(-x-R)dx$$

$$\to h\{F(+\infty) - F(-\infty)\},$$

as $R \to +\infty$, and the limit is finite by hypothesis.

We note that $G(x) \to 0$, as $|x| \to \infty$, (by definition and by the hypothesis on F), and by partial integration,

$$\lim_{R\to\infty}\left[\int_{-R}^{R} e^{ixy}dG(y)\right] = \lim_{R\to\infty}\left[e^{ixy}G(y)\right]_{y=-R}^{y=R} - ix \lim_{R\to\infty}\int_{-R}^{R} e^{ixy}G(y)dy,$$

which gives

$$\int_{-\infty}^{\infty} e^{ixy}dG(y) = -ix\int_{-\infty}^{\infty} e^{ixy}G(y)dy,$$

so that, by (1.7), we have

$$\varphi(x)\left[\frac{e^{-ihx}-1}{-ix}\right] = \int_{-\infty}^{\infty} e^{ixy}G(y)dy.$$

Since $F(x) = \frac{1}{2}[F(x+0) + F(x-0)]$, this implies (by Theorem 5, Ch.I), that

$$G(x) = \frac{1}{2\pi}\lim_{R\to\infty}\int_{-R}^{R}\varphi(y)\,\frac{e^{-ihy}-1}{-iy}\,e^{-ixy}dy,$$

for each x. For $x = 0$, this gives (1.5).

<u>*Theorem 2*</u>. *Let* F *be of bounded variation in* $(-\infty,+\infty)$, <u>*and*</u> *let* $F(x) = \frac{1}{2}[F(x+0) + F(x-0)]$, *for all* x, *and let* $\varphi(x) = \int_{-\infty}^{\infty} e^{ixy}dF(y)$. *Then we have*

$$(1.8) \qquad \int_0^x \{F(y) - F(-y)\}dy = \frac{1}{\pi}\int_{-\infty}^{\infty}\varphi(y)\,\frac{1-\cos xy}{y^2}\,dy,$$

the integral on the right-hand side converging absolutely (since φ *is bounded).*

<u>Proof</u>. We have, for $R > 0$,

$$\frac{1}{R\pi}\int_{-\infty}^{\infty}\varphi(y)\,\frac{(1-\cos Ry)}{y^2}\,dy = \frac{1}{\pi}\int_{-\infty}^{\infty}dF(y)\int_{-\infty}^{\infty}e^{ixy}\,\frac{1-\cos Ry}{Ry^2}\,dy$$

$$= \int_{-R}^{R} \left(1 - \frac{|x|}{R}\right) dF(x) = \frac{1}{R} \int_{0}^{R} [F(y) - F(-y)]dy.$$

(see Ex.2, §1, Ch.I; also Ex.1,§8, Ch.I).

Remarks. Let $F_1(x)$, $F_2(x)$,..., be *non-decreasing*, and *bounded* functions in $(-\infty < x < \infty)$, and let

$$\varphi_n(x) = \int_{-\infty}^{\infty} e^{ixy} dF_n(y), \qquad n = 1,2,\ldots . \tag{1.9}$$

The following examples show the difficulty in preserving the equality sign in (1.9) after letting $n \to +\infty$.

(1.10) Let $F_n(x) = 0$, for $x < n$; $F_n(x) = 1$, for $x \geq n$. Then $F_n(x) \to 0$, for each x, as $n \to +\infty$, but $\varphi_n(x) = e^{inx}$ does *not* tend to a limit as $n \to +\infty$, except for $x = 2\pi k$, where k is an integer.

(1.11) Let $F_n(x)$ be continuous, $F_n(x) = 0$ for $x \leq -n$, $F_n(x) = 1$ for $x \geq n$, $F_n(x) = \frac{x}{2n} + \frac{1}{2}$ for $-n < x < n$ (linear). Then $F_n(x) \to \frac{1}{2}$, for each x, as $n \to \infty$, and

$$\varphi_n(x) = \int_{-n}^{n} \frac{e^{ixy}}{2n} dy = \frac{\sin nx}{nx},$$

so that $\varphi_n(x) \to 0$, as $n \to \infty$, for each $x \neq 0$, and $\varphi_n(x) \to 1$, at $x = 0$, as $n \to \infty$, so that

$$\lim_{n\to\infty} \varphi_n(x) \neq \int_{-\infty}^{\infty} e^{ixy} d(\lim_{n\to\infty} F_n(y)).$$

§2. Distribution functions, and characteristic functions

In order to avoid some of the difficulties pointed up by the above examples, we shall now consider *non-decreasing functions* $F(x)$, which have finite limits at $x = \pm\infty$. We suppose further that

$$F(-\infty) = 0, \qquad F(+\infty) = 1. \tag{2.1}$$

We call such a function F a *distribution* function. We call the Fourier-

Stieltjes transform of such an F, namely

$$\varphi(x) = \int_{-\infty}^{\infty} e^{ixy} dF(y),$$

a *characteristic function*, corresponding to the given distribution function F.

It is clear that φ is *continuous* and *bounded*. (Note that $|\varphi(x)| \leq F(+\infty) - F(-\infty) = 1$, and $\varphi(0) = 1$). It is also *hermitian*, in the sense that $\overline{\varphi(-x)} = \varphi(x)$.

Theorem 1 implies that given φ, the distribution function F is defined uniquely at the points of continuity of F, because $F(-\infty) = 0$, $F(+\infty) = 1$. The next theorem shows how the convergence of a sequence of distribution functions implies the convergence of the corresponding sequence of characteristic functions.

<u>*Theorem 3*</u>. *Let* (F_n) *be a sequence of distribution functions, and* (φ_n) *the corresponding sequence of characteristic functions, so that*

$$\varphi_n(x) = \int_{-\infty}^{\infty} e^{ixy} dF_n(y). \tag{2.2}$$

If F_n *converges to a distribution function* F *as* $n \to \infty$, *at the points of continuity of* F, *and if* φ *is the characteristic function of* F, *then*

$$\varphi_n(x) \to \varphi(x) \quad \text{as } n \to \infty, \tag{2.3}$$

and the convergence is uniform in <u>*every*</u> *finite interval.*

<u>Proof</u>. We note that the set of discontinuities of F is countable. Given $\varepsilon > 0$, we choose $R > 0$, such that $F(x)$ is continuous both at $x = R$ and $x = -R$, with $F(-R) < \varepsilon$, $F(R) > 1-\varepsilon$, and such that $F_n(\pm R) \to F(\pm R)$, as $n \to \infty$. Then we have, for $n \geq n_0$, also

$$F_n(-R) < \varepsilon, \quad F_n(R) > 1-\varepsilon.$$

Now

$$\varphi(x)-\varphi_n(x) = \int_{|y|\leq R} e^{ixy} d\{F(y)-F_n(y)\} + \int_{|y|>R} e^{ixy} dF(y)$$

$$- \int_{|y|>R} e^{ixy} dF_n(y) = I_1 + I_2 + I_3, \text{ say.}$$

Then we have

$$|I_2| \le (1 - F(R) + F(-R)) < 2\varepsilon, \quad \text{for all } x,$$

while

$$|I_3| \le (1 - F_n(R) + F_n(-R)) < 2\varepsilon, \text{ for } n \ge n_0, \text{ and all } x.$$

On the other hand,

$$I_1 = \Big[(F(y) - F_n(y))e^{ixy}\Big]_{y=-R}^{y=+R} - ix \int_{|y|\le R} \{F(y) - F_n(y)\}e^{ixy}dy.$$

Since $F_n(\pm R) \to F(\pm R)$, as $n \to \infty$, the first term on the right-hand side tends to zero, *uniformly in x*. And

$$\left|-ix \int_{|y|\le R} \{F(y) - F_n(y)\}e^{ixy}dy\right| \le |x| \int_{|y|\le R} |F(y) - F_n(y)|dy \to 0,$$

uniformly in each *finite* x-interval since $|F| \le 1$, $|F_n| \le 1$; hence so does I_1. Thus we have altogether

$$|\varphi(x) - \varphi_n(x)| < \varepsilon + 2\varepsilon + 2\varepsilon, \text{ for } n \ge n_0,$$

and for x in any *finite* interval.

The next result is a kind of converse to Theorem 3.

<u>Theorem 4</u>. *If $\varphi_n(x) \to \varphi(x)$, for <u>each</u> x, as $n \to \infty$, where φ_n is the characteristic function corresponding to the distribution function F_n, <u>and</u> if φ is continuous at the origin, then F_n converges to a distribution function F, at the points of continuity of F, and φ is the characteristic function of F.*

<u>Proof</u>. Since $F_n(x)$ is a non-decreasing function of x for $n = 1,2,\ldots,$ and since $0 \le F_n(x) \le 1$, for all $x \in (-\infty,\infty)$, there exists (by Helly's theorem) a subsequence F_{n_k} of F_n of non-decreasing functions (of x) which converges, as $n_k \to \infty$, to a non-decreasing function F(x). Clearly $0 \le F(x) \le 1$. We shall see that F *is* a distribution function, and that $F_n \to F$, at *every point of continuity* of F.

By definition, we have

$$\varphi_n(x) = \int_{-\infty}^{\infty} e^{ixy}dF_n(y).$$

If we use formula (1.8), we get, for $R > 0$,

$$\int_0^R \{F_{n_k}(y) - F_{n_k}(-y)\}dy = \frac{1}{\pi}\int_{-\infty}^{\infty} \varphi_{n_k}(y)\,\frac{1-\cos Ry}{y^2}dy. \tag{2.4}$$

Since $|\varphi_n(x)| \le 1$, for all $x \in (-\infty,\infty)$, and $\varphi_n(x) \to \varphi(x)$, for each x, as $n\to\infty$, we get on letting $n_k \to \infty$ in (2.4),

$$\int_0^R [F(y) - F(-y)]dy = \frac{1}{\pi}\int_{-\infty}^{\infty} \varphi(y)\,\frac{1-\cos Ry}{y^2}\,dy,$$

or

$$\frac{1}{R}\int_0^R [F(y) - F(-y)]dy = \frac{1}{2\pi}\int_{-\infty}^{\infty} \varphi(y)H_R(y)dy, \quad H_R(y) = \frac{\sin^2(Ry/2)}{R(y/2)^2}. \tag{2.5}$$

Since $\varphi_n(0) = 1$ for $n = 1,2,\ldots$, and $\varphi(x) = \lim\limits_{n\to\infty}\varphi_n(x)$, we have $\varphi(0) = 1$, and since $\int_{-\infty}^{\infty}\left(\frac{\sin\alpha}{\alpha}\right)^2 d\alpha = \pi$, we have $\frac{1}{2\pi}\int_{-\infty}^{\infty} H_R(y)dy = 1$. (cf. Ex.1, Th.13, Ch.I).Hence, for a suitably chosen $\delta > 0$, we have

$$\frac{1}{2\pi}\int_{-\infty}^{\infty}\varphi(y)H_R(y)dy - 1 = \frac{1}{2\pi}\int_{-\infty}^{\infty}\{\varphi(y)-\varphi(0)\}H_R(y)dy$$

$$= \frac{1}{2\pi}\left(\int_{|y|<\delta} + \int_{|y|\ge\delta}\right)\{\varphi(y)-\varphi(0)\}H_R(y)dy,$$

$$= I_1 + I_2, \text{ say.}$$

Because of the continuity of φ at the origin, given $\varepsilon > 0$, there exists a $\delta > 0$, such that

$$|I_1| \le \sup_{|y|<\delta}|\varphi(y)-\varphi(0)|\cdot 1 < \varepsilon,$$

while

$$|I_2| \le 2\,\frac{1}{2\pi}\int_{|y|\ge\delta} R\left[\frac{\sin(Ry/2)}{(Ry)/2}\right]^2 dy = \frac{1}{\pi}\int_{|t|\ge R\delta}\left(\frac{\sin t/2}{t/2}\right)^2 dt \to 0,$$

as $R\to\infty$ for a suitably chosen, but fixed, $\delta > 0$. Hence

$$\lim_{R\to\infty}\frac{1}{2\pi}\int_{-\infty}^{\infty}\varphi(y)H_R(y)dy = 1,$$

which implies, because of (2.5), that

$$\lim_{R\to\infty}\frac{1}{R}\int_0^R \{F(y)-F(-y)\}dy = 1. \tag{2.6}$$

The left-hand side equals $F(+\infty) - F(-\infty)$. Since $0 \leq F(x) \leq 1$, *and* F is non-decreasing, we deduce that $F(+\infty) = 1$, $F(-\infty) = 0$. Hence F is a distribution function. By Theorem 3, we have

$$\lim_{n_k\to\infty} \varphi_{n_k}(x) = \varphi(x) = \int_{-\infty}^{\infty} e^{ixy}dF(y).$$

If there exists a point of *continuity* x_0 of F, such that $F_n(x_0) \nrightarrow F(x_0)$ then we can find a subsequence $F_{n'_k}$, which converges *everywhere* to a distribution function F^*, *and* such that $F^*(x_0) \neq F(x_0)$. If, for example, $F^*(x_0) > F(x_0)$, then since F is continuous at x_0, and F^* is non-decreasing, we have $F^*(x) > F(x)$, for $x \in (x_0, x_0+\eta)$, for some $\eta > 0$, sufficiently small. (Similarly if $F^*(x_0) < F(x_0)$, then $F^*(x) < F(x)$ for $x \in (x_0-\eta', x_0)$ for some $\eta' > 0$). This is impossible since, by Theorem 3, φ is also the characteristic function of F^*, and the characteristic function determines the distribution function up to an additive constant (Th.1) and $F^*(-\infty) = F(-\infty) = 0$.

§3. Positive definite functions: the theorems of Bochner and of F. Riesz

We shall consider classes of functions f which can be represented as Fourier-Stieltjes transforms.

Lemma 1. If f(t) *is a complex-valued function which is measurable and finite for* $-\infty < t < \infty$, *and satisfies the condition*

$$\sum_{\mu=1}^{m} \sum_{\nu=1}^{m} f(t_\mu - t_\nu)\rho_\mu\bar{\rho}_\nu \geq 0 \tag{3.1}$$

for any integer $m \geq 1$, *and arbitrary real numbers* $t_1, t_2, \ldots, t_m$ *and arbitrary complex numbers* $\rho_1, \rho_2, \ldots, \rho_m$, *then we have*

$$f(0) \geq 0; \quad f(-t) = \overline{f(t)}; \quad |f(t)| \leq f(0). \tag{3.2}$$

Proof. On taking m = 1 in (3.1), we obtain

$$|\rho_1|^2 f(0) \geq 0,$$

so that $f(0) \geq 0$.

On taking $m = 2$, $t_1 = t$, $t_2 = 0$, we obtain

$$(3.3) \qquad f(0)[|\rho_1|^2 + |\rho_2|^2] + f(t)\rho_1\overline{\rho}_2 + f(-t)\overline{\rho}_1\rho_2 \geq 0.$$

If we choose $\rho_2 = 1$ in (3.3), we get

$$f(0)[|\rho_1|^2+1] + f(t)\rho_1 + f(-t)\overline{\rho}_1 \geq 0,$$

and if in this we set $\rho_1 = 1, i$ respectively, we see that $f(t) + f(-t)$ is real (since $f(0)$ is real), *and* that $f(t) - f(-t)$ is purely imaginary, hence $f(-t) = \overline{f(t)}$.

If $f(0) = 0$, $\rho_1 = 1$, $\rho_2 = -f(t)$, then (3,3) gives $-2|f(t)|^2 \geq 0$, since $f(-t) = \overline{f(t)}$, hence $f(t) = 0$, for all t.

If $f(0) > 0$, $\rho_1 = f(0)$, $\rho_2 = -f(t)$, then (3.3) gives

$$f(0)[|f(0)|^2 + |f(t)|^2] \geq f(0)\{2|f(t)|^2\},$$

which implies that $|f(t)| \leq f(0)$, and (3.2) is proved.

Lemma 2 *(F. Riesz). For any complex-valued measurable function* f, *condition (3.1) implies that*

$$(3.4) \qquad \int_{-\infty}^{\infty} \int_{-\infty}^{\infty} f(t-s)\rho(t)\overline{\rho}(s)ds\, dt \geq 0,$$

for any $\rho \in L_1(-\infty,\infty)$, *provided that* $f(0)$ *is finite.*

Proof. If $A > 0$, and $\rho \in L_2(-A,A)$, we take $\rho_\mu = \rho(t_\mu)$ in (3.1) and integrate with respect to t_μ, $\mu = 1,2,\ldots,m$, where $m > 1$. We get from *each* of the diagonal terms (in which $\mu = \nu$)

$$(2A)^{m-1} f(0) \int_{-A}^{A} |\rho(t)|^2 dt,$$

while *each* of the remaining terms gives

$$(2A)^{m-2} \int_{-A}^{A} \int_{-A}^{A} f(t-s)\rho(t)\overline{\rho}(s)ds\, dt.$$

Hence we have

$$m(2A)^{m-1} f(0) \int_{-A}^{A} |\rho(t)|^2 dt + m(m-1)(2A)^{m-2} \int_{-A}^{A} \int_{-A}^{A} f(t-s)\rho(t)\overline{\rho}(s)ds\, dt \geq 0 .$$

On dividing throughout by $m(m-1)(2A)^{m-2}$, and then letting $m \to \infty$, we get

$$(3.5) \qquad \int_{-A}^{A} \int_{-A}^{A} f(t-s)\rho(t)\overline{\rho}(s)\,ds\,dt \geq 0,$$

for any $\rho \in L_2(-A,A)$. If $\rho \in L_1(-A,A)$, (3.5) still holds good, for if we define, for any integer $n \geq 1$,

$$\rho_n(t) = \begin{cases} \rho(t), & \text{if } |\rho(t)| \leq n, \\ n, & \text{if } |\rho(t)| > n, \end{cases}$$

then we have $|\rho_n(t)| \leq |\rho(t)| \in L_1(-A,A)$, and $|\rho_n(t)| \leq n$, for all $t \in (-A,A)$. Hence $\rho_n \in L_2(-A,A) \cdot \cap \cdot L_1(-A,A)$, so that (3.5) holds with $\rho_n(t)$ in place of $\rho(t)$. Since $\rho_n(t) \to \rho(t)$, as $n \to \infty$, and $|f(t)| \leq f(0)$, *and* $f(0)$ is *finite*, we obtain (3.5) for any $\rho \in L_1(-A,A)$. If we then let $A \to \infty$, we obtain (3.4).

Lemma 3. If a measurable function f *(as in Lemma 1) satisfies (3.1), and* $f(0)$ *is finite, then for any* $\varepsilon > 0$, *the function*

$$(3.6) \qquad f_\varepsilon(x) = e^{-\varepsilon x^2} f(x), \qquad -\infty < x < \infty,\ \varepsilon > 0,$$

also satisfies (3.1); and $f_\varepsilon(x) \in L_1(-\infty < x < \infty)$.

Proof. We have

$$\sum_{\mu=1}^{m} \sum_{\nu=1}^{m} f_\varepsilon(t_\mu - t_\nu)\rho_\mu\overline{\rho}_\nu$$

$$= \sum_{\mu=1}^{m} \sum_{\nu=1}^{m} \rho_\mu\overline{\rho}_\nu\ f(t_\mu - t_\nu) e^{-\varepsilon(t_\mu - t_\nu)^2}$$

$$(3.7) \qquad = \sum_{\mu=1}^{m} \sum_{\nu=1}^{m} \rho_\mu\overline{\rho}_\nu\ f(t_\mu - t_\nu) \frac{1}{2(\pi\varepsilon)^{1/2}} \int_{-\infty}^{\infty} e^{-t^2/4\varepsilon + i(t_\mu - t_\nu)t}\,dt,$$

if we note that

$$\int_{-\infty}^{\infty} e^{-x^2 + i\alpha x}\,dx = \sqrt{\pi} \cdot e^{-\alpha^2/4}, \qquad \text{(see §1, Ch.I)}$$

and replace x by $x/(2\sqrt{\varepsilon})$, and α by $2\alpha\sqrt{\varepsilon}$.

Now the expression in (3.7) equals the integral

$$\int_{-\infty}^{\infty} e^{-t^2/4\varepsilon} \left(\sum_{\mu=1}^{m} \sum_{\nu=1}^{m} \left(\rho_\mu\, e^{it_\mu t}\right)\overline{\left(\rho_\nu\, e^{it_\nu t}\right)}\, f(t_\mu - t_\nu) \right) dt \geq 0,$$

since f satisfies (3.1) by hypothesis. It follows from (3.7) that f_ε satisfies (3.1). Since f(0) is finite, and (by Lemma 1), $|f(t)| \leq f(0)$, we note that f is bounded, hence $f_\varepsilon \in L_1(-\infty,\infty)$ for every $\varepsilon > 0$.

Lemma 4. *Suppose that* f *is a complex-valued measurable function on* $(-\infty,\infty)$ *which satisfies (3.1), with* f(0) *finite. Let* $f_\varepsilon(x) = e^{-\varepsilon x^2} f(x)$, $\varepsilon > 0$, *so that (by Lemma 3)* $f_\varepsilon \in L_1(-\infty,\infty)$. *Then we have*

$$\hat{f}_\varepsilon(\alpha) \geq 0, \quad \hat{f}_\varepsilon \in L_1(-\infty,\infty), \quad -\infty < \alpha < \infty,$$

and

$$f_\varepsilon(x) = \frac{1}{2\pi} \int_{-\infty}^{\infty} \hat{f}_\varepsilon(\alpha) e^{-i\alpha x} d\alpha, \tag{3.8}$$

for almost all $x \in (-\infty,\infty)$. *In particular, (3.8) holds at every point of continuity* x *of* f_ε, *and therefore of* f.

Proof. By Lemma 2, f satisfies condition (3.4). In it we choose ρ, such that

$$\rho(t) = e^{-2\varepsilon t^2} e^{i\alpha t}, \quad -\infty < t < \infty,$$

where α is *real* and *fixed*. Then (3.4) becomes

$$\int_{-\infty}^{\infty} \int_{-\infty}^{\infty} f(x-y) e^{-2\varepsilon(x^2+y^2)}\, e^{i\alpha(x-y)} dx\, dy \geq 0.$$

If we make the substitutions

$$x - y = u, \; x + y = v,$$

then we have

$$\frac{1}{2}\int_{-\infty}^{\infty}\int_{-\infty}^{\infty} f(u)e^{-\varepsilon(u^2+v^2)}\, e^{iu\alpha}du\, dv$$

$$= \frac{1}{2}\int_{-\infty}^{\infty}\left(\int_{-\infty}^{\infty} e^{-\varepsilon v^2}dv\right) e^{-\varepsilon u^2} f(u)e^{iu\alpha}du$$

$$= \frac{1}{2}\int_{-\infty}^{\infty}\left(\int_{-\infty}^{\infty} e^{-\varepsilon v^2}dv\right) f_\varepsilon(u)e^{iu\alpha}du \geq 0.$$

Now $f_\varepsilon \in L_1(-\infty,\infty)$, and f_ε is bounded on $(-\infty,\infty)$, since $f(0)$ is finite, and the last inequality shows that f_ε has a positive Fourier transform. By Theorem 12 of Chapter I, $\hat{f}_\varepsilon \in L_1(-\infty,\infty)$ and (3.8) follows.

Lemma 5. Let

$$K(\alpha) = \begin{cases} 1-|\alpha|, & \text{for } |\alpha| \leq 1, \\ 0, & \text{for } |\alpha| > 1, \end{cases}$$

so that $K \in L_1(-\infty,\infty)$, *and let* $\hat{K} = H$. *If, for* $R > 0$, $K_R(\alpha) \equiv K(\frac{\alpha}{R})$, *and* $H_R(\alpha) = RH(R\alpha)$, *then*

$$\hat{K}_R = H_R \quad \textit{and} \quad \hat{H}_R = 2\pi\, K_R.$$

(If $f \in L_1(-\infty,\infty)$, $\hat{f}(\alpha) = \int_{-\infty}^{\infty} f(x)e^{i\alpha x}dx$, *as in Ch.I, (8.1)). If* f_ε *is defined for every* $\varepsilon > 0$ *as in Lemma 3, then we have*

$$\int_{-\infty}^{\infty} H_R(x-y)f_\varepsilon(y)dy = \int_{-\infty}^{\infty} K_R(\alpha)e^{i\alpha x}\,\hat{f}_\varepsilon(-\alpha)d\alpha. \tag{3.9}$$

Proof. Since K is even, H is even, so that the left-hand side of (3.9) equals

$$\int_{-\infty}^{\infty} H_R(y-x)f_\varepsilon(y)dy = \int_{-\infty}^{\infty} f_\varepsilon(x+t)H_R(t)dt.$$

By the composition rule (Ch.I, (1.13)), this equals

$$\int_{-\infty}^{\infty} K_R(\alpha)e^{-i\alpha x}\,\hat{f}_\varepsilon(\alpha)d\alpha = \int_{-\infty}^{\infty} K_R(\alpha)e^{+i\alpha x}\,\hat{f}_\varepsilon(-\alpha)d\alpha,$$

since $f_\varepsilon \in L_1(-\infty,\infty)$ by Lemma 4, $K_R \in L_1(-\infty,\infty)$, $H_R \in L_1(-\infty,\infty)$, K_R is even, and $\hat{K}_R = H_R$.

Lemma 6. Let f *be a complex-valued measurable function defined on* $(-\infty,\infty)$, *which satisfies condition (3.1), and is continuous for* $x = 0$. *Let*

(3.10) $$f_n(x) = e^{-x^2/n}f(x),\ n \geq 1,\ n \text{ integral},\ -\infty < x < \infty.$$

Then there exists a non-decreasing, bounded function $V_n(t)$, $-\infty < t < \infty$, *such that*

(3.11) $$\frac{1}{2\pi}\int_{-\infty}^{\infty} H_R(x-y)f_n(y)\,dy = \int_{-\infty}^{\infty} K_R(\alpha)e^{i\alpha x}dV_n(\alpha),\ -\infty < x < \infty.$$

Proof. If f is continuous at x = 0, then f(0) is finite, and by Lemma 4, $\hat{f}_n(-\alpha) \geq 0$, for $-\infty < \alpha < \infty$, $\hat{f}_n \in L_1(-\infty,\infty)$. If we *define*

(3.12) $$V_n(t) = \frac{1}{2\pi}\int_{-\infty}^{t} \hat{f}_n(-\alpha)\,d\alpha,\quad -\infty < t < \infty,$$

then $V_n(t) \geq 0$, and $V_n(t)$ is a non-decreasing function of t for $n = 1,2,\ldots$. If f is continuous at the origin, then, by Lemma 4, we have

$$f_n(0) = \frac{1}{2\pi}\int_{-\infty}^{\infty} \hat{f}_n(-\alpha)\,d\alpha,$$

so that

(3.13) $$0 \leq V_n(t) \leq f_n(0) = f(0),\ \text{for } -\infty < t < \infty,\ n = 1,2,\ldots .$$

Now (3.9) gives (3.11).

Theorem 5. *Let* f *be a complex-valued measurable function defined on* $(-\infty,\infty)$, *which is continuous at the origin, and which satisfies condition (3.1). Then there exists a non-decreasing, bounded function* V(t), *such that*

(3.14) $$f(x) = \int_{-\infty}^{\infty} e^{ixt}dV(t),\quad -\infty < x < \infty$$

for almost all $x \in (-\infty,\infty)$. *If* f *is continuous everywhere, then (3.14) holds for every* $x \in (-\infty,\infty)$.

Proof. We have to let $n \to \infty$ in relation (3.11) of Lemma 6 just proved, and then let $R \to \infty$. Since $\{V_n(t)\}$ is uniformly bounded, by (3.13), by a theorem of Helly there exists a subsequence (n_k), and a non-decreasing function V, such that $|V(t)| \leq f(0)$, and $V_{n_k} \to V$ pointwise on $(-\infty,\infty)$.

Since $K_R(\alpha)$ in Lemma 5 vanishes for $|\alpha| \geq R$, and is continuous, we have, by a theorem of Helly-Bray (see the Notes)

$$(3.15)\qquad \lim_{k\to\infty}\int_{-\infty}^{\infty} K_R(\alpha)e^{i\alpha x}dV_{n_k}(\alpha) = \int_{-\infty}^{\infty} K_R(\alpha)e^{i\alpha x}dV(\alpha),$$

for each $R > 0$, on the right-hand side of (3.11). If we take the left-hand side of (3.11), and let $n \to \infty$, through the same subsequence (n_k) as in (3.15), we obtain, since $|f_n(y)| \le f_n(0) = f(0)$, and $f(0)$ is finite, and $H_R \in L_1(-\infty,\infty)$,

$$(3.16)\qquad \lim_{k\to\infty}\frac{1}{2\pi}\int_{-\infty}^{\infty} H_R(x-y)f_{n_k}(y)dy = \frac{1}{2\pi}\int_{-\infty}^{\infty} H_R(x-y)f(y)dy,$$

by Lebesgue's theorem on dominated convergence. From (3.16) and (3.15), and (3.11), we thus obtain

$$(3.17)\qquad \frac{1}{2\pi}\int_{-\infty}^{\infty} H_R(x-y)f(y)dy = \int_{-\infty}^{\infty} K_R(\alpha)e^{i\alpha x}dV(\alpha), \quad (R > 0).$$

We now let $R \to \infty$. The function $K_R(\alpha)e^{i\alpha x}$ is continuous in α, and vanishes outside $(-R,R)$, and $K_R(\alpha) \to 1$, as $R \to \infty$, for each α, while

$$\int_{-\infty}^{\infty} |dV(\alpha)| \le V(+\infty) - V(-\infty) \le f(0) < \infty.$$

Hence

$$(3.18)\qquad \lim_{R\to\infty}\int_{-\infty}^{\infty} K_R(\alpha)e^{i\alpha x}dV(\alpha) = \int_{-\infty}^{\infty} e^{i\alpha x}dV(\alpha),$$

on the right-hand side of (3.17). We shall show that the left-hand side of (3.17) gives

$$(3.19)\qquad \lim_{R\to\infty}\frac{1}{2\pi}\int_{-\infty}^{\infty} f(y)H_R(x-y)dy = \lim_{R\to\infty}\frac{1}{2\pi}(f*H_R)(x) = f(x),$$

for almost all $x \in (-\infty,\infty)$, so that (3.14) follows, and the theorem is proved.

Let $\omega > 0$, arbitrarily chosen and kept fixed, and let $f_\omega(y) = f(y)$, for $|y| < \omega$; while $f_\omega(y) = 0$, for $|y| \ge \omega$. Since f is bounded and measurable in $(-\omega,\omega)$, $f \in L_1(-\omega,\omega)$; hence $f_\omega \in L_1(-\infty,\infty)$. By Theorem 10 of Chapter I, we have

$$\lim_{R\to\infty}\frac{1}{2\pi}(f_\omega * H_R)(x) = f_\omega(x), \text{ for almost all } x \in (-\infty,\infty).$$

That is to say that

$$(3.20)\qquad \lim_{R\to\infty} \frac{1}{2\pi} \int_{-\omega}^{\omega} f(y)H_R(x-y)\,dy = f(x),$$

for almost all $x \in [-\omega,\omega]$. By a further conclusion in Theorem 10 of Chapter I, this relation holds for *every* $x \in (-\omega,\omega)$, if f is everywhere continuous.

On the other hand, if $|x| < \omega$, we have

$$(3.21)\qquad \left|\frac{1}{2\pi} \int_{|y|\geq\omega} f(y)H_R(x-y)\,dy\right| \leq \frac{f(0)}{2\pi\cdot R} \int_{|y|\geq\omega} \frac{\sin^2[R(x-y)/2]}{[(x-y)/2]^2}\,dy$$

$$\to 0, \text{ as } R\to\infty.$$

Thus (3.21) and (3.20) lead to the relation

$$(3.19)\qquad \lim_{R\to\infty} \frac{1}{2\pi} (f*H_R)(x) = f(x),$$

for *almost all* $x \in [-\omega,\omega]$, where $\omega > 0$ is *arbitrary*; and for *every* $x \in (-\omega,\omega)$ if f is continuous everywhere. This combined with (3.18) and (3.17) proves the theorem.

The next result is a kind of converse and easier to prove.

Theorem 6 (Bochner). Let V(t) *be a non-decreasing function of* t, $-\infty < t < \infty$, *which is bounded everywhere. Then the function*

$$(3.22)\qquad f(x) = \int_{-\infty}^{\infty} e^{ixt}\,dV(t), \quad -\infty < x < \infty,$$

is continuous, and bounded, and satisfies condition (3.1).

Proof. As already remarked in §1, f is bounded and continuous. To verify (3.1) we have only to note that

$$\sum_{\mu=1}^{m}\sum_{\nu=1}^{m} f(t_\mu-t_\nu)\rho_\mu\bar{\rho}_\nu = \sum_{\mu=1}^{m}\sum_{\nu=1}^{m} \rho_\mu\bar{\rho}_\nu \int_{-\infty}^{\infty} e^{it(t_\mu-t_\nu)}\,dV(t)$$

$$= \int_{-\infty}^{\infty} \left(\sum_{\mu=1}^{m} \rho_\mu e^{it_\mu t}\right)\left(\sum_{\nu=1}^{m} \bar{\rho}_\nu e^{-it_\nu t}\right)dV(t)$$

$$= \int_{-\infty}^{\infty} \left|\sum_{\mu=1}^{m} \rho_\mu e^{it_\mu t}\right|^2 dV(t) \geq 0,$$

since V is non-decreasing, and the integrand is positive.

(3.23) Definition. A complex-valued function f(x) defined for $-\infty < x < \infty$ is said to be positive definite, if it is continuous, and bounded, and satisfies condition (3.1).

Theorems 5 and 6 yield the following

Theorem 7 (Bochner). In order that a function f(x) *defined on* $(-\infty,\infty)$ *may be written as*

$$f(x) = \int_{-\infty}^{\infty} e^{ixt}dV(t),$$

where V *is a non-decreasing, bounded function on* $(-\infty,\infty)$, *it is necessary and sufficient that* f *be positive definite.*

Since a positive definite function is, by definition, continuous in $(-\infty,\infty)$, the proof of Theorem 7 does *not* require the full force of Theorem 5. The following theorem is sufficient.

Theorem 8. If f *is a complex-valued function defined on* $(-\infty,\infty)$, *which is continuous, and satisfies condition (3.1), then*

$$f(x) = \int_{-\infty}^{\infty} e^{ixt}dV(t), \tag{3.24}$$

where V *is a non-decreasing, bounded function of* $t \in (-\infty,\infty)$.

Proof. Define for each integer n, $n \geq 1$,

$$f_n(x) = e^{-x^2/n}f(x), \quad -\infty < x < \infty.$$

Then we have $f_n \in L_1(-\infty,\infty)$, for every n, since f is bounded, and by Lemma 4, with $\varepsilon = 1/n$, we have $\hat{f}_n \in L_1(-\infty,\infty)$, and

$$f_n(x) = \frac{1}{2\pi}\int_{-\infty}^{\infty} \hat{f}_n(\alpha)e^{-i\alpha x}d\alpha = \frac{1}{2\pi}\int_{-\infty}^{\infty} \hat{f}_n(-\alpha)e^{i\alpha x}d\alpha, \tag{3.25}$$

where $\hat{f}_n(-\alpha) \geq 0$, for $-\infty < \alpha < \infty$, for *every* $x \in (-\infty,\infty)$, since f is continuous (see Th.11', Ch.I, p. 45).

If we define, as in (3.12),

$$V_n(t) = \frac{1}{2\pi}\int_{-\infty}^{t} \hat{f}_n(-\alpha)\,d\alpha, \quad -\infty < t < \infty,$$

then (3.25) becomes

$$(3.26) \qquad f_n(x) = \int_{-\infty}^{\infty} e^{i\alpha x} dV_n(\alpha).$$

Now $V_n(-\infty) = 0$, while $V_n(+\infty) = f_n(0) = f(0)$.

We consider two cases

Case (i). Let $f(0) = 1$. Then V_n is a *distribution function*, for each n, with f_n as its *characteristic function*. Since $f_n(x) \to f(x)$, as $n \to \infty$, for each x, and f is continuous (at the origin), Theorem 4 implies that V_n converges to a distribution function V (at the points of continuity of V), whose characteristic function is f. That is to say, (3.26) implies that

$$f(x) = \int_{-\infty}^{\infty} e^{i\alpha x} dV(\alpha)$$

as claimed.

Case (ii). Let $f(0) \neq 1$. Condition (3.1) implies that $f(0) \geq 0$. If $f(0) = 0$, then f is identically zero, and V is a constant. If $f(0) \neq 0$, then $f(0) > 0$, and the function $g(x) = \frac{f(x)}{f(0)}$ is continuous, and satisfies condition (3.1), with $g(0) = 1$. By what has been proved in Case (i), it follows that

$$f(x) = \int_{-\infty}^{\infty} e^{i\alpha x} d(f(0)\cdot V(\alpha)) = \int_{-\infty}^{\infty} e^{i\alpha x} dV_1(\alpha),$$

where $V_1(\alpha) = f(0)\cdot V(\alpha)$, and V_1 is non-decreasing, and bounded.

The above proof yields the following

(3.27) *Corollary. Positive definite functions* f, *with the property* $f(0) = 1$, *are characteristic functions.*

An easy extension of Theorem 7 is the following

Theorem 9 (Bochner). Let $g \in L_2(-\infty,\infty)$, *and let*

(3.28) $$f(x) = \int_{-\infty}^{\infty} g(x+t)\overline{g(t)}dt.$$

Then we have

$$f(x) = \int_{-\infty}^{\infty} e^{ix\alpha}dV(\alpha),$$

where V *is non-decreasing, and bounded, in* $(-\infty,\infty)$.

Proof. Since

$$|f(x)|^2 \leq \int_{-\infty}^{\infty} |g(x+t)|^2 dt \int_{-\infty}^{\infty} |g(t)|^2 dt = (f(0))^2,$$

which is finite, since $\|g\|_2$ is finite, f is *bounded*.

We also have, for any fixed x,

$$|f(x)-f(x_1)|^2 \leq \int_{-\infty}^{\infty} |g(x+t)-g(x_1+t)|^2 dt \int_{-\infty}^{\infty} |g(t)|^2 dt,$$

and if we let $x_1 \to x$, then the right-hand side tends to zero, so that f is *continuous*.

Since

$$f(x-y) = \int_{-\infty}^{\infty} g(x-y+t)\,\overline{g}(t)dt = \int_{-\infty}^{\infty} g(x+t)\overline{g(y+t)}dt,$$

we have

$$\sum_{\mu=1}^{m} \sum_{\nu=1}^{m} f(x_\mu - x_\nu)\rho_\mu\overline{\rho}_\nu = \int_{-\infty}^{\infty} \left|\sum_{\mu=1}^{m} g(x_\mu+t)\rho_\mu\right|^2 dt \geq 0.$$

It follows that f is positive definite, and hence, by Theorem 7, that

$$f(x) = \int_{-\infty}^{\infty} e^{ix\alpha}dV(\alpha).$$

§4. A uniqueness theorem

For a special kind of Fourier transform, namely

$$\int_{-\infty}^{\infty} e^{ixy}\varphi(y)dy,$$

where $\varphi \in L_1(-\omega,\omega)$, for each *finite* $\omega > 0$, we shall prove a theorem of uniqueness, which generalizes Theorem 7 of Chapter I.

Theorem 10 (Offord). If $\varphi(u) \in L_1(-\omega < u < \omega)$, *for every finite* $\omega > 0$, *and if*

$$\lim_{\omega\to\infty} \int_{-\omega}^{\omega} \varphi(u)e^{iux}du = 0, \tag{4.1}$$

for every real x, *then* $\varphi(u) = 0$, *for almost every* $u \in (-\infty,\infty)$.

We first prove three preliminary lemmas.

Lemma 1 (Schwarz). If $f(x)$ *is real-valued and continuous for* $a \le x \le b$, $b > a$, *and*

$$\lim_{h\to 0} \frac{f(x+h) + f(x-h) - 2f(x)}{h^2} = 0, \tag{4.2}$$

for every $x \in [a,b]$, *then* f *is linear in* (a,b); *that is to say,* $f(x) = Ax+B$, *where* A *and* B *are some constants.*

Proof. Let

$$F(x) = \theta\left[f(x) - f(a) - \frac{x-a}{b-a}\{f(b) - f(a)\}\right],$$

where $\theta = \pm 1$, and let

$$G(x) = F(x) - \tfrac{1}{2}\varepsilon(x-a)(b-x),$$

where $\varepsilon > 0$. Then G is a continuous function of x for $x \in [a,b]$, with $G(a) = G(b) = 0$; so is F, with $F(a) = F(b) = 0$.

Case (i). If $F(x) = 0$, for every $x \in [a,b]$, the lemma is immediate since then we have

$$f(x) = \frac{f(b) - f(a)}{b-a}\cdot x + \frac{bf(a) - af(b)}{b-a} .$$

Case (ii). If there exists a point c in (a,b), such that $F(c) \neq 0$, choose θ in such a way that $F(c) > 0$, and choose ε so small that $G(c) > 0$. Since G is continuous in $[a,b]$, it attains its maximum M, say, at a point $x_0 \in (a,b)$, and $M > 0$, since $G(c) > 0$. We have, because of (4.2),

$$\lim_{h\to 0} \frac{G(x_0+h) + G(x_0-h) - 2G(x_0)}{h^2} = \varepsilon > 0.$$

But $G(x_0+h) \le G(x_0)$, and $G(x_0-h) \le G(x_0)$, so that the above limit is negative or zero, which implies a contradiction. Hence $F(x) = 0$ for all $x \in [a,b]$, and the lemma follows.

Lemma 2. Let $\varphi(u) \in L_1(-\omega < u < \omega)$, *for each finite* $\omega > 0$, *and let*

$$(4.3) \qquad \lim_{\omega\to\infty} I(\omega) \equiv \lim_{\omega\to\infty} \int_{-\omega}^{\omega} \varphi(u)du = 0.$$

If $\varphi(u) \to 0$, *as* $|u| \to \infty$, *then*

$$(4.4) \qquad \int_{-\infty}^{\infty} \varphi(u) \left(\frac{\sin uh}{uh}\right)^2 du$$

exists for every $h > 0$, *and tends to zero as* $h \downarrow 0$.

Proof. Since $\int_{-\omega}^{\omega} \varphi(u)du = \int_0^{\omega} \Phi(u)du$, where $\Phi(u) = \varphi(u) + \varphi(-u)$, assumption (4.3) implies that $I(\omega) = \int_0^{\omega} \Phi(u)du \to 0$, as $\omega \to \infty$. The integral in (4.4) equals

$$\int_0^{\infty} \left(\frac{\sin uh}{uh}\right)^2 dI(u),$$

and this, in turn, equals, by partial integration,

$$\left[I(u) \left(\frac{\sin uh}{uh}\right)^2\right]_{u=0}^{\infty} - \int_0^{\infty} I(u) \frac{d}{du}\left[\left(\frac{\sin uh}{uh}\right)^2\right]du$$

$$= -\int_0^{\infty} I\left(\frac{u}{h}\right) H'(u)du, \quad \text{where } H' \in L_1(0,\infty),\ H(u) = \left(\frac{\sin u}{u}\right)^2 .$$

Since $I(\frac{u}{h})$ is bounded, we have by Lebesgue's theorem on dominated convergence,

$$\lim_{h\downarrow 0} \int_0^{\infty} I\left(\frac{u}{h}\right) H'(u)du = \int_0^{\infty} \lim_{h\downarrow 0} I\left(\frac{u}{h}\right) H'(u)du = 0.$$

Lemma 3. Let $\varphi(u) \in L_1(-\omega < u < \omega)$, *for each finite* $\omega > 0$; *let*

$$\lim_{\omega\to\infty} \int_{-\omega}^{\omega} \varphi(u)e^{iux}du = 0,$$

for every real x, *and* $\varphi(u) \to 0$, *as* $|u| \to \infty$. *Then* $\varphi(u) = 0$ *for almost all* $u \in (-\infty,\infty)$.

Proof. We take, as we may, $\varphi(u)e^{iux}$ instead of $\varphi(x)$ in Lemma 2. Then we have

$$\int_{-\infty}^{\infty} \varphi(u)e^{iux} \left(\frac{\sin uh}{uh}\right)^2 du \to 0, \text{ as } h \downarrow 0. \tag{4.5}$$

Define

$$F(x) = \int_{-\infty}^{\infty} \varphi(u) \frac{e^{iux} - L(x,u)}{-u^2} du,$$

where

$$L(x,u) = \begin{cases} 1+iux, & \text{for } |u| < 1, \\ 0, & \text{for } |u| \geq 1. \end{cases}$$

For a fixed $h \neq 0$, consider the function

$$\frac{F(x+h)+F(x-h)-2F(x)}{h^2} = \int_{-\infty}^{\infty} \frac{\varphi(u)}{(iuh)^2} \left\{e^{i(x+h)u} + e^{i(x-h)u} - 2\,e^{iux}\right\}du$$

$$= \int_{-\infty}^{\infty} \frac{\varphi(u)}{(iuh)^2} e^{iux} \left\{e^{\frac{iuh}{2}} - e^{\frac{-iuh}{2}}\right\}^2 du$$

$$= \int_{-\infty}^{\infty} \varphi(u)\, e^{iux} \left(\frac{\sin(\frac{uh}{2})}{(uh/2)}\right)^2 du. \tag{4.6}$$

Because of (4.5), and Lemma 1, it follows that F is *linear*, hence $F(x+h) + F(x-h) - 2F(x) \equiv 0$, which implies, in turn, that

$$\int_{-\infty}^{\infty} \varphi(u)e^{iux} \left(\frac{\sin \frac{uh}{2}}{uh/2}\right)^2 du \equiv 0.$$

But the integral on the left-hand side is the Fourier transform of a function belonging to $L_1(-\infty,\infty)$. Hence by Theorem 7 of Chapter I, $\varphi(u)$ is zero for almost all $u \in (-\infty,\infty)$.

We remark that this method of proof goes back to Riemann in his treatment of the uniqueness of trigonometric series.

Proof of Theorem 10.

Assumption (4.1) holds with y+x, and y-x, in place of x (where y is real), so that it implies that

$$\int_{-\omega}^{\omega} \varphi(u)e^{iuy} \cos ux \cdot dx \to 0, \text{ as } \omega \to \infty, \tag{4.7}$$

for *every* real y and x. If we set, for fixed y,

$$\Psi(u) = \varphi(u)e^{iuy} + \varphi(-u)e^{-iuy},$$

then (4.7) becomes

$$(4.8) \qquad \int_0^{\omega} \Psi(u) \cos ux\cdot du \to 0, \text{ as } \omega \to \infty,$$

for every real x. Now let

$$\Psi_1(v) = \int_0^v \Psi(u)\,du = \int_{-v}^{v} \varphi(u)e^{iuy}du, \quad v > 0.$$

Then (4.8) implies, in particular, that

$$(4.9) \qquad \Psi_1(v) \to 0, \text{ as } v \to \infty.$$

Since

$$\int_0^{\omega} \Psi(u) \cos ux\cdot du = \Psi_1(\omega) \cos \omega x - \Psi_1(0) + x \int_0^{\omega} \Psi_1(u) \sin ux\cdot du,$$

where $\Psi_1(0) = 0$, (4.9) implies that

$$x \int_0^{\omega} \Psi_1(u) \sin ux\cdot du \to 0, \text{ as } \omega \to \infty,$$

which implies that for every $x \neq 0$, we have

$$\int_0^{\omega} \Psi_1(u) \sin ux\cdot du \to 0, \text{ as } \omega \to \infty,$$

where $\Psi_1(u) \to 0$, as $u \to \infty$, by (4.9). By Lemma 3, $\Psi_1(u) = 0$, for almost all $u \in (-\infty,\infty)$. Since Ψ_1 is absolutely continuous, it follows that $\Psi(u)$ is zero for almost all u, hence $\varphi(u)e^{iuy} + \varphi(-u)e^{-iuy} = 0$, for almost all u. On setting $y = 0$, we have: $\varphi(u) + \varphi(-u) = 0$, or $\varphi(u) = -\varphi(-u)$, for almost all u. Hence

$$\Psi(u) = 2i\ \varphi(u) \sin uy,$$

and therefore $\varphi(u) = 0$ for almost all $u \in (-\infty,\infty)$.

<u>Remark</u>. If $\varphi(u) = e^{iu^2}$, then $\varphi(u) \notin L_1(-\infty < u < \infty)$, although

$$\lim_{\omega\to\infty} \int_{-\omega}^{\omega} \varphi(u)e^{iux}du$$

is finite for all $x \in (-\infty,\infty)$. (See Ex.10, Step (iii), §1, Ch.I).

Another such example is provided by

$$\varphi(u) = \exp(\alpha u + ie^{u}), \quad 0 < \alpha < 1.$$

Notes

Chapter I

§1. Theorem 1 is a generalization of the basic result on Fourier series, which states that the Fourier coefficients of an *integrable* function tend to zero, which was proved by Riemann for Riemann-integrable functions, and extended by Lebesgue to Lebesgue-integrable functions. The idea of the second proof sketched in the Remarks following Theorem 1 is due to Lebesgue, *Bull. Sociéte Math. de France*, 38(1910), 184-210.

A peculiar generalization of Theorem 1 has been given by Bochner and Chandrasekharan in [1], Th.46, Ch.III. See the later definition of *pseudo-characters* by Bochner [5] Ch.3, p.53.

A characteristic function in the sense used here is *not* the same as in Chapter III, hence the alternative term "indicator function".

For the rôle of Hermite functions in Fourier analysis, see, for instance, N. Wiener [3].

The standard work on Bessel functions is Watson's [1]. A short introduction is given in Ch.XVII of Whittaker and Watson [1].

Example 10 is due to S. Ramanujan, J. Indian Math. Soc. 11 (1919), 81-87; Coll. Papers, No. 23 [1]. The integral in Step (i) is evaluated by Cauchy's theorem in Lindelöf's book [1]; see p. 49 for the motivation of the proof.

§2. For the notion of an algebra, and basic facts about algebras, see, for instance, G. Birkhoff and S. MacLane: "A survey of modern algebra", p. 225.

For an introduction to Fourier analysis on groups, see the classic by Weil [1]; also Loomis [1], Naimark [1], the Appendix in Goldberg [1], Rudin [1], and Reiter [1], where the contributions of A. Beurling, I.E. Segal, and others, are described.

§3. For the theory of distributions, in general, with applications, see the classics by L. Schwartz [1], and I.M. Gelfand and G.E. Shilov [1]. For distributions in connexion with Fourier transforms, in particular, see Ch.I of Hörmander's book [1], also Yosida [1], Ch.VI, and Donoghue [1].

The function ω in (3.12) was introduced by Wiener [1], p.562.

§4. The localization theorem here is motivated by the one on Fourier series due to Riemann, see Hardy and Rogosinski [1], pp. 39-42, and Zygmund [1], Ch.II, §6, p. 53, §8.

The examples of Mellin transforms given here are of frequent occurrence in analytic number theory, see, for instance, the author's book [3]. Several more are given by Titchmarsh [3].

§5. For Poisson's summation formula, see Bochner [1]; his proof is also given in the author's book [3]. For some special applications see, for instance, Zygmund [1], Vol.I, Ch.II, §13.

For the theta-relation (5.8), in the general setting of theta-functions, see, for instance, the author's book [4], where the connexion with elliptic functions, and the theory of numbers, is elucidated on an elementary level.

There is also an L_2-version in one variable, see Boas [2].

§6. The proof of the uniqueness in Theorem 7 (without the use of summability and general inversion) can be effected by the use of a piece-wise linear (trapezoidal) function instead of the function $\omega_{c,\varepsilon}$; see Bochner and Chandrasekharan [1], Ch.I, §6, Th.5.

For a sharper version of the uniqueness theorem due to A.C. Offord [1], see Th.10, Ch.III.

§7. The motivation for the summability theorems here is again supplied by the theory of Fourier series, see, for instance, Hardy and Rogosinski [1], Ch.V, p. 70; Bochner and Chandrasekharan [1], §7.

For properties (7.6) and (7.20) of integrable functions, and for the definition of the "Lebesgue set", see, for instance, Titchmarsh [2], §11.6.

Convolution integrals of the type (7.1) are of importance in the theory of approximation. See Butzer and Nessel [1], where generalized singular integrals of the type of Cauchy-Poisson, Gauss-Weierstrass, Fejér, and Bochner-Riesz, are dealt with in detail. For more general methods, see Stein [3], Stein and Weiss [1].

§8. Example 3, following Theorem 11', is used by C.L. Siegel in his proof of Hamburger's theorem on the Riemann zeta-function. See, for instance, the author's book [3], Ch.II, §5.

Theorem 12 is due to Bochner and Chandrasekharan [1], Ch.I, Th.9, p.20; also p.211, where it is commented upon. A generalization was later given by Bochner in his book [5], p.25, Th.2.2.1.

Theorems 13 and 14 make it possible to define the Fourier transform on $L_2(-\infty,\infty)$ and prove Plancherel's theorem [cf. Ch.II] by starting from the subspace $L_1 \cdot \cap \cdot L_2$.

§9. The systematic use of summability in norm seems to have originated with Wiener. For Theorem 17 see Bochner and Chandrasekharan [1], Ch.I, §10, who also proved further results in that direction.

On Weyl's form of the Riesz-Fischer theorem, see Weyl [1], and Wiener's [3] remarks; also Stone's [1], p.26; and J. von Neumann's [1], p. 109-111.

§10. For the central limit theorem, see Cramér [2], Ch.17, §4, and Feller [1], Ch.VIII, §4. According to Cramér, the theorem was first stated by Laplace in 1812; a rigorous proof under "fairly general" conditions was given by Liapounoff; and the problem of finding the most general conditions of validity was solved by Feller, Khintchine, and Lévy. The proof given here differs only in detail from that given, for instance, by Dym and McKean [1], Ch.2, §7.

§11. Theorem 21 is the analogue, for Fourier transforms, of a classical theorem on the absolute convergence of Fourier series due to N. Wiener [2] and P. Lévy [1]. See Zygmund [1], Vol.I, Ch.VI, §5. The proof given here differs only in detail from that of R.R. Goldberg [1], Ch.2, §9, which is itself closely modelled on Bochner's proof [2] of the Wiener-Lévy theorem.

§12. Wiener was the first to study "closure" properties of functions in $L_1(-\infty,\infty)$ and in $L_2(-\infty,\infty)$, and relate them to Fourier transform theory. See Wiener [2]. The proof given here of Theorem 23 is the same as Bochner's [2]. An algebraic reformulation of the theorem would be that every proper closed ideal of $L_1(-\infty,\infty)$ is contained in a maximal ideal. The problem of characterizing the sub-class of functions f in $L_1(-\infty,\infty)$ which have the property that $\overline{S}_f$ is the intersection of the maximal ideals containing it has received attention. For the work of Beurling and others, see Pollard [1], and Reiter [1]. For generalizations of Theorem 23, see Ch.4 of Goldberg [1], Reiter [1], where further references can be found, e.g. to Agmon and Mandelbrojt [1], Malliavin [1], and others.

§13. Theorems 24 and 25 are due to Wiener [2]. Theorem 26 is due to Littlewood [1], and forms the prototype for many of the tauberian theorems of the period before Wiener. Littlewood's theorem can be proved directly, and simply, as Karamata [1] has shown, by the use of Weierstrass's theorem on the approximation of continuous functions, instead of Wiener's theorem on the L_1-closure. See Wiener's own remarks [3], and Wielandt's [1] arrangement of Karamata's proof.

Wiener's work on tauberian theorems has been carried forward notably by H.R. Pitt [1].

Albert Stadler [1] has recently proved a tauberian theorem, *with* remainder, of the Wiener-Ikehara type (see, for instance, the author's book [2]), which yields the more refined forms of the prime number theorem as corollaries. See Wiener's remarks [2], p.93, on this possibility.

Bochner and Chandrasekharan [1], Th.29, p.54, subsume Karamata's theorem as part of another theorem which characterizes what they call the *Karamata extension* of the kernel $e^{-\alpha}$. The nature of this extension in the case of general kernels seems not to be known.

§14. Theorems 27 and 28 are due to Bochner and Chandrasekharan [1], Ch.I, §15. See Butzer and Nessel [1] Ch.7, for later developments. Equations (14.1) and (14.2) arise in connexion with the problem of conduction of heat, see Carslaw [1], §§16,45.

§15. The proof of (15.24) given here is the same as the one given by Bochner in his book [1]; he comments that according to Burkhardt [1] pp.1165-1173, the cases k = 2,3 are due to Poisson and Cauchy, and that the "theorem is also not new for k arbitrary". A second proof is given by Bochner and Chandrasekharan [1], pp.71-74.

The introduction of the spherical mean $f_x(t)$ is due to Bochner. He carried the idea further into the study of multiple Fourier series. See Bochner [4], followed by Chandrasekharan [1], Chandrasekharan and Minakshisundaram [1], and [2], Ch.IV, and H. Joris [1]. Important work with quite different techniques has been accomplished on topics in multiple Fourier series by E.M. Stein, and others. See, for instance, Ch.VII of Stein and Weiss [1].

The evaluation of $V_k(s)$ by induction is done, for instance, by Walfisz [1], p.41.

Chapter II

§2, §3. Different proofs of Plancherel's [1] theorem have been given by Titchmarsh [2], Bochner [1], F. Riesz [1], Wiener [3], M.H. Stone [1], p.104, and Bochner and Chandrasekharan [1]. For the Remarks following (2.29) see Stein and Weiss [1], p.18. For (3.5) see Bochner and Chandrasekharan [1], p.99 .

§4. Theorem 6 is due to Wiener; for Theorem 7 see Bochner and Chandrasekharan [1], Ch.IV, §10.

§5. Weyl's proof of the inequality under somewhat stronger hypotheses is given in Appendix I to his book [2]. His proof in the second edition differs in detail from the one given in the first. The proof given here differs only in detail from the one given in their book by Dym and McKean [1].

§6. The Phragmén-Lindelöf [1] principle takes many forms. See, for instance, Calderón-Zygmund [1], Littlewood [2], p.107, Titchmarsh [2],

§5.71. For Hardy's theorem, see Hardy [1], and Titchmarsh [3], p.174, where further references are given.

§7. Paley and Wiener [1] were the first to make a systematic study of Fourier transforms in the complex domain (one variable). The proof given here differs only in detail from the one presented by Dym and McKean [1], Ch.3. For functions of exponential type see, for instance, Boas [1].

§8. For generalities on Fourier orthogonal series see Kaczmarz and Steinhaus [1], Ch.II, where several examples of orthogonal systems are given, including Rademacher's [1], and Walsh's [1] which can be defined, after Paley [1], in terms of Rademacher's functions. Let $\chi_0(t) = 1$, and if N is a positive integer, expressed in the binary scale as $N = 2^{n_1} + 2^{n_2} + \ldots$, with $n_1 > n_2 > \ldots$, then $\chi_N(t) = \varphi_{n_1}(t) \cdot \varphi_{n_2}(t) \ldots$, where the (φ_n) are Rademacher's functions. The system (χ_n) is orthonormal over (0,1), and complete, and is known as Walsh's.

Bessel's inequality (in several variables) has been shown by Siegel [1] to yield Minkowski's first theorem on lattice points in convex sets [cf. the author's book [2], p.99]. Atle Selberg [1] has shown that Bombieri's large sieve inequality can be viewed as a form of Bessel's inequality in a Hilbert space, cf. H.L. Montgomery [1].

For a concise introduction to Fourier trigonometric series in $L_2(0,2\pi)$ see Hardy and Rogosinski [1]. Series and integrals can be treated together on a group space. See Butzer and Nessel [1]. For Lebesgue's proof of the completeness of the trigonometric system, see Hardy and Rogosinski [1], or Zygmund [1], Vol.I, Ch.I, §6.

§9. Hardy's [2] interpolation formula is also treated in Zygmund [1], Vol.II, p.276. As Dr. Albert Stadler has remarked, the condition of boundedness in Theorem 12 can be replaced by one of polynomial growth, in which case formula (9.7) will assume a more general form.

§10. S. Bernstein's work [1] is also presented in Zygmund [1], Vol. II, p.11, Ch.X, and p.276, Ch.XVI. Zygmund's inequality for the integrated derivative of a trigonometric polynomial, as a generalization of Theorem 14, is given by him immediately after Bernstein's result. N.G. de Bruijn [1] has given generalizations of Bernstein's theorem for polynomials in the complex domain. For the Remark preceding Theorem 14, see Siegel [1]. See also Stein [2].

§11. For the extension of the Paley-Wiener theorem to k dimensions, $k > 1$, see Plancherel and Pólya [1], Stein [2], Stein and Weiss [1], Ch.III, Th.4.9. The last-mentioned reference connects the theorem with the analysis of H^p-spaces. See Narasimhan [1], Ch.3, for the rôle of the Fourier transform in analytical problems on manifolds; also Ehrenpreis [1] in connexion with several complex variables.

Chapter III

§1. For the basic theory of Stieltjes integrals see, for instance, Burkill and Burkill [1], Ch.6, and Widder [1], Ch.I.

§2. As Zygmund has remarked, the essence of Theorems 3 and 4 is a classical result of the calculus of probability, in a form strengthened by Cramér. See Zygmund [1], Vol.II, Ch.XVI, Th.(4.24), p.262. See also Cramér [2], Ch.10. Bochner has a generalization to E_k, see Th.3.2.1 of his book [5], p.56.

For Helly's theorem used in the proof of Theorem 4, see, for instance, Widder [1], Ch.I, §16, Th.16.2.

§3. Theorems 6 and 7 are due to Bochner, see [1], Th.23. He refers to previous work by F. Bernstein and M. Mathias. The generalization, without the assumption of continuity, is due to F. Riesz [1], who uses for the proof, however, his theorem on the representation of positive linear functionals, which is *not* used here in the proof of Theorem 5. For the Helly-Bray theorem used, see, for instance, Widder [1], p.31, Th.16.4. It is not necessarily true when the interval of integration is infinite, as Widder makes clear, hence the introduction of the kernel $K_R(x)$. Carleman [1], p.98, gives a proof of Bochner's theorem *using* the Poisson integral representation of functions which are positive and harmonic in a half-plane. A proof of the latter (see, for instance, Verblunsky [1]) can be obtained by using Herglotz's theorem [1] on the representation of positive, harmonic functions in a circle (which is stated, for instance, in Stone [1], p.571), or more directly, as has been done by Loomis and Widder [1] using the theorems of Helly, and of Helly-Bray. It should be remarked, however, that all these representation theorems are, more or less, of the same order of difficulty as Bochner's theorem, or Stone's spectral theorem [1], p.331, as was early recognized by F. Riesz.

Apropos Corollary (3.27), see Cramér [1]. For Theorem 9 see Bochner [3], p.329. Bochner has also a generalization to E_k, [5], Theorem 3.2.3, p.58.

For a generalization to distributions, see Schwartz [1], Vol.II, p.132, Th.XVIII; Schwartz makes a reference to Weil [1], p.122.

§4. Lemma 1 is due to H.A. Schwarz. It is quoted by G. Cantor, *J. für Math.* 72 (1870), 141; and is given by Schwarz himself in his *Ges. Abhandlungen*, II (1890), 341-343, with a reference to Cantor's quotation. Prof. Raghavan Narasimhan has remarked that a rearrangement of Schwarz's argument is better adapted to generalization. "If f is real-valued, and continuous on (a,b), and $\limsup_{h\to 0} (\Delta_h^2 f)(x) = 0$, for all $x \in (a,b)$, where $\Delta_h^2 f(x) = h^{-2}\{f(x+h)+f(x-h)-2f(x)\}$, then f is linear. To prove this, it is sufficient to prove that if $\limsup_{h\to 0} \Delta_h^2 f \geq 0$, then f is convex (i.e. if $\ell(x) = cx+d$, and $f(\alpha) \leq \ell(\alpha)$, $f(\beta) \leq \ell(\beta)$, where $a < \alpha < \beta < b$, then $f(x) \leq \ell(x)$ for $\alpha \leq x \leq \beta$), since one can argue similarly with $\liminf_{h\to 0} \Delta_h^2(-f)$. Since $\Delta_h^2 \ell = 0$ for any linear function ℓ, it is enough to prove that if $\limsup_{h\to 0} \Delta_h^2 f \geq 0$, and $f(\alpha) \leq 0$, $f(\beta) \leq 0$, then $f(x) \leq 0$ on $[\alpha,\beta]$, i.e. that f has no maximum on (α,β). Replacing f(x) by $f(x) + \varepsilon x^2$, $\varepsilon > 0$, one has only to show that if $\limsup_{h\to 0} \Delta_h^2 f > 0$, on (a,b), then f has no local maximum on (a,b). But this is obvious, for if x_0 is a local maximum, then $\Delta_h^2 f(x_0) \leq 0$ for h small enough, since $f(x_0) \geq f(x_0+h)$, $f(x_0) \geq f(x_0-h)$." Cf. Narasimhan [2], p.21-25.

Theorem 10 is due to A.C. Offord [1], and is the integral analogue of Cantor's fundamental theorem on the uniqueness of trigonometric series, which asserts that if a trigonometric series converges everywhere to zero, it vanishes identically; all its coefficients are zero. See, for instance, Zygmund [1], Vol.I, Ch.IX, p.326. Offord also proved [2] a stronger theorem in which the hypothesis of convergence of the integral in (4.1) is replaced by $(C,1)$ summability. Offord shows that the stronger theorem is a "best possible", in the sense that even one exceptional point cannot be permitted, and $(C,1)$ summability cannot be relaxed to $(C,1+\varepsilon)$ summability for any $\varepsilon > 0$. Zygmund's proof [1], Vol.

II, Ch.XVI, §10, of Offord's first theorem is based on an equiconvergence theorem for trigonometric integrals and series which he treats in Vol.II, Ch.XVI, §9, and on results from Riemann's theory of trigonometric series which he treats in Vol.I, Ch.IX. For the use of equiconvergence theorems in analytic number theory, see, for instance, the author's book [3], Ch.VIII. A generalization of Offord's stronger theorem to several variables would be of interest, though perhaps not easy. An equiconvergence theorem for trigonometric integrals in two variables has been given by H. Keller [2]. Functions of bounded variation in two variables come into play, and it is a moot question whether the notion of Vitali variation could be replaced by that of Frechet, as Morse [1] and Transue did in another context.

References

Agmon, S., and Mandelbrojt, S. [1] Acta. Sci. Math. Szeged, 12 (1950), 167-176.

Bernstein, S. [1] Mém. Acad. Roy. Belgique, 2me série, 4 (1912), 1-104.

Beurling, A. [1] Neuvième Congrès des mathématiciens scandinaves, (Helsingfors, 1938), 345-366.

Boas, R.P. [1] Entire functions (New York, 1954)
[2] J. London Math. Soc. 21 (1946), 102-105.

Bochner, S. [1] Vorlesungen über Fouriersche Integrale (Leipzig, 1932; Chelsea, 1948)
[2] Lectures on Fourier analysis (Princeton University, 1936; Ann Arbor, 1937)
[3] Lectures on Fourier integrals, Annals of Math. Studies, No. 42, (Princeton, 1959)
[4] Trans. American Math. Soc. 40 (1936), 175-207.
[5] Harmonic analysis and the theory of probability (Univ. California, 1955)

Bochner, S., and Chandrasekharan, K. [1] Fourier transforms, Annals of Math. Studies, No. 19 (Princeton, 1949)

Burkhardt, H. [1] Trigonometrische Reihen und Integrale (bis etwa 1850), Enzyklopädie der Math. Wiss. II, 1, (ii) Analysis (Teubner, 1904-1916, pp. 819-1354)

Burkill, J.C., and Burkill, H. [1] A second course in mathematical analysis (Cambridge, 1970)

Butzer, P.L., and Nessel, R.J. [1] Fourier analysis and approximation, I, (Birkhäuser, 1971)

Calderôn, A.P., and Zygmund, A. [1] Contributions to Fourier analysis, Annals of Math. Studies, No. 25 (Princeton, 1950), 166-188.

Campbell, G.A., and Foster, R.M. [1] Fourier integrals for practical applications, Bell Telephone System Tech. Publications (USA, 1931)

Carleman, T. [1] L'integrale de Fourier et questions qui s'y rattachent (Almqvist and Wiksell, 1944)

Carslaw, H.S. [1] Mathematical theory of conduction of heat in solids, 2nd edn. (London, 1921)

Chandrasekharan, K. [1] Proc. London Math. Soc. (2) 50 (1948), 210-229.
[2] Introduction to analytic number theory (Springer, 1968)
[3] Arithmetical functions (Springer, 1970)
[4] Elliptic functions (Springer, 1985)

Chandrasekharan, K., and Minakshisundaram, S. [1] Duke Math. J. 14 (1947), 731-753.
[2] Typical means (Oxford, 1952)

Cramér, H. [1] Trans. Amer. Math. Soc. 46 (1939), 190-201.
[2] Mathematical methods of statistics (Princeton, 1946)

De Bruijn, N.G. [1] Indagationes Math. 9 (1947), 591-598.

Donoghue, W.F. Jr. [1] Distributions and Fourier transforms, (Academic Press, 1969)

Dym, H., and McKean, H.P. [1] Fourier series and integrals (Academic Press, 1972)

Ehrenpreis, L. [1] Fourier analysis in several complex variables, (Wiley-Interscience, 1970)

Feller, W. [1] An introduction to probability theory and its applications, I,II, (John Wiley, 1950, 1966)

Gelfand, I.M., and Shilov, G.E. [1] Generalized functions, Vol.1, (Academic Press, 1964)

Goldberg, R.R. [1] Fourier transforms, Cambridge Tracts, No.52 (1961)

Hardy, G.H. [1] J. London Math. Soc. 8 (1933), 227-231.
[2] Proc. Cambridge Phil Soc. 37 (1941), 331-348.

Hardy, G.H., Littlewood, J.E., and Pólya, G. [1] Inequalities (Cambridge, 1934)

Hardy, G.H., and Rogosinski, W.W. [1] Fourier series, Cambridge Tracts, No. 38, (1944)

Herglotz, G. [1] Ber. Verhandl. Sächs. Gesellschaft Wiss. Leipzig, Math.-Phys. Kl. 63 (1911), 501-511.

Hörmander, L. [1] Linear partial differential operators (Springer, 1963)

Joris, H. [1] Math. Zeitschrift, 103 (1968), 61-66.

Kaczmarz, S., and Steinhaus, H. [1] Theorie der Orthogonalreihen, (Warsaw, 1935)

Karamata, J. [1] Math. Zeitschrift, 32 (1930), 319-20; 33 (1931), 294-300; J für Math. 164 (1931), 27-40.

Katznelson, Y. [1] An introduction to harmonic analysis (John Wiley, 1968)

Keller, H. [1] Dissertation Nr. 5725, ETH Zürich (1976)

Lévy, P. [1] Compositio Math. 1 (1934), 1-14.

Lindelöf, E. [1] Le Calcul des Résidus (Paris, 1905; Chelsea, 1947)

Littlewood, J.E. [1] Proc. London Math. Soc. 9 (1910), 434-448.
[2] Lectures on the theory of functions (Oxford, 1944)

Loomis, L.H. [1] An introduction to abstract harmonic analysis (van Nostrand, 1953)

Loomis, L.H., and Widder, D.V. [1] Duke Math. J. 9 (1942), 643-645.

Malliavin, P. [1] Publ. Math. IHES, 2 (1959), 61-68.

Montgomery, H.L. [1] Bull. Amer. Math. Soc. 84 (1978), p. 559.

Morse, M. and Transue, W. [1] Contributions to Fourier analysis, Annals of Math. Studies, No. 25 (1950), 46-103.

Naimark, M.A. [1] Normed rings (P. Noordhoff, 1964)

Narasimhan, Raghavan [1] Analysis on real and complex manifolds (Elsevier, 1968; third printing, 1985)
[2] Complex analysis in one variable, (Birkhäuser, 1985)

Neumann, J. von [1] Math. Annalen, 102 (1929), 109-111.

Offord, A.C. [1] J. London Math. Soc. 11 (1936), 171-174.
[2] Proc. London Math. Soc. (2) 42 (1937), 422-480.

Paley, R.E.A.C. [1] Proc. London Math. Soc. 34 (1932), 241-279.

Paley, R.E.A.C., and Wiener, N. [1] Fourier transforms in the complex domain, Amer. Math. Soc. Colloq. Publ. XIX (1934)

Phragmén, E., and Lindelöf, E. [1] Acta Math. 31 (1908), 381-406.

Pitt, H.R. [1] Tauberian theorems (Oxford, 1958)

Plancherel, M. [1] Rendi. di Palermo, 30 (1910), 289-335.

Plancherel, M., and Pólya, G. [1] Comm. Math. Helv. 9 (1937), 224-248.

Pollard, H. [1] Duke Math. J. 20 (1953), 499-512.

Rademacher, H. [1] Math. Annalen, 87 (1922), 112-138.

Ramanujan, S. [1] J. Indian Math. Soc. 11 (1919), 81-87; Collected Papers (Cambridge, 1927; Chelsea, 1962), No. 23

Reiter, H. [1] Classical harmonic analysis and locally compact groups (Oxford, 1968)

Riemann, B. [1] Gesammelte Werke, 2. Auflage, Leipzig (1892), 227-271.

Riesz, F. [1] Acta Sci. Math. Szeged, 3 (1927), 235-241.
[2] Acta. Sci. Math. Szeged, 6 (1933), 184-198.

Rudin, W. [1] Fourier analysis on groups (Interscience, 1962)

Schwartz, L. [1] Théorie des distributions, I, II (Hermann, Paris, 1950-51)
[2] Mathematics for the physical sciences (Hermann, Paris, 1966)

Seeley, R.T. [1] An introduction to Fourier series and integrals (Benjamin, 1966)

Selberg, A. [1] Collected papers, II (Springer-Verlag, to appear)

Siegel, C.L. [1] Acta Math. 65 (1935), 307-323.

Sneddon, I.N. [1] Fourier transforms (McGraw-Hill, 1951)

Stadler, Albert [1] Dissertation Nr. 8073, ETH Zürich (1986); to appear in Comm. Math. Helvetici

Stein, E.M. [1] Trans. Amer. Math. Soc. 83 (1956), 482-492.
[2] Annals of Math. 65 (1957), 582-592.
[3] Singular integrals and differentiability properties of functions (Princeton, 1970)

Stein, E.M., and Weiss, G. [1] Introduction to Fourier analysis on Euclidean spaces (Princeton, 1971)

Stone, M.H. [1] Linear transformations in Hilbert space and their applications to analysis, Amer. Math. Soc. Colloq. Publ. XV (1932)

Titchmarsh, E.C. [1] Proc. London Math. Soc. (2) 23 (1924), 279-289.
[2] The theory of functions, 2nd edn. (Oxford, 1939)
[3] Introduction to the theory of Fourier integrals (Oxford, 1937; 2nd edn. 1948)

Verblunsky, S. [1] Proc. Cambridge Phil. Soc. 31 (1935), 482-507.

Walfisz, A. [1] Gitterpunkte in mehrdimensionalen Kugeln (Warsaw, 1957)

Walsh, J.L. [1] Amer. J. Math. 55 (1923), 5-24.

Watson, G.N. [1] A treatise on the theory of Bessel functions (Cambridge 1922)

Weil, A. [1] L'integration dans les Groupes topologiques et ses Applications (Hermann, Paris, 1940)

Weyl, H. [1] Math. Annalen, 67 (1909), 225-245.
[2] The theory of groups and quantum mechanics (German Edn., Hirzel, 1928; English Edn., Dover Reprint, New York, 1949)

Whittaker, E.T., and Watson, G.N. [1] A course of modern analysis (Cambridge, 1902; 4th edn. 1927)

Widder, D.V. [1] The Laplace transform (Princeton, 1941)

Wielandt, H. [1] Math. Zeit. 56 (1952) 206-207.

Wiener, N. [1] Math. Annalen, 95 (1926), 557-584.
[2] Annals of Math. 33 (1930), 1-100.
[3] The Fourier integral and certain of its applications (Cambridge, 1933)
[4] Fourier series and integrals (MIT Lectures, 1936-37)

Yosida, K. [1] Functional analysis (Springer, 1965)

Zygmund, A. [1] Trigonometric series, I, II, (Cambridge, 1959)

K. T. Smith

Power Series from a Computational Point of View

Universitext

1987. 2 figures. VIII, 132 pages.
ISBN 3-540-96516-5

Contents: Taylor Polynomials. – Sequences and Series. – Power Series and Complex Differentiability. – Local Analytic Functions. – Analytical Continuation. – Index.

The purpose of this book is to explain the use of power series in performing concrete calculations, such as approximating definite integrals or solutions to differential equations. This focus may seem narrow but, in fact, such computations require the understanding and use of many of the important theorems of elementary analytic function theory, for example Cauchy's Integral Theorem, Cauchy's Inequalities, and Analytic Continuation and the Monodromy Theorem. These computations provide an effective motivation for learning the theorems, and a sound basis for understanding them.

Springer-Verlag Berlin
Heidelberg New York London
Paris Tokyo Hong Kong